T0289827

Cognitive Radio Networks

Efficient Resource Allocation
in Cooperative Sensing, Cellular
Communications, High-Speed
Vehicles, and Smart Grid

OTHER COMMUNICATIONS BOOKS FROM AUERBACH

Cognitive Radio Networks

Efficient Resource Allocation
in Cooperative Sensing, Cellular
Communications, High-Speed
Vehicles, and Smart Grid

Tao Jiang
Zhiqiang Wang
Yang Cao

CRC Press
Taylor & Francis Group
Boca Raton   London   New York

CRC Press is an imprint of the
Taylor & Francis Group, an **informa** business

CRC Press
Taylor & Francis Group
6000 Broken Sound Parkway NW, Suite 300
Boca Raton, FL 33487-2742

Printed on acid-free paper
Version Date: 20150129

International Standard Book Number-13: 978-1-4987-2113-4 (Hardback)

Visit the Taylor & Francis Web site at
http://www.taylorandfrancis.com

and the CRC Press Web site at
http://www.crcpress.com

Contents

Preface ix

Acknowledgments xi

About the Authors xiii

1 Introduction 1
    1.1 Cognitive Radio-Based Networks 1
    1.2 Opportunistic Spectrum Access Networks 4
    1.3 Cognitive Radio Networks with Cooperative Sensing 6
    1.4 Cognitive Radio Networks for Cellular Communications 8
    1.5 Cognitive Radio Networks for High-Speed Vehicles 10
    1.6 Cognitive Radio Networks for a Smart Grid 11
    1.7 Content and Organization 13

**2 Transmission Slot Allocation in an Opportunistic Spectrum Access
Network** 15
    2.1 Single-User Single-Channel System Model 16
    2.2 Probabilistic Slot Allocation Scheme 18
    2.3 Optimal Probabilistic Slot Allocation 21
        2.3.1 Baseline Performance 21
        2.3.2 Exponential Distribution 21
        2.3.3 Hyper-Erlang Distribution 22
    2.4 Performance Analysis and Evaluation 23
        2.4.1 Impact of Sensing Errors 23
        2.4.2 Impact of Unknown Primary User Idle Period Distribution . 26
        2.4.3 Performance Comparisons 27
    2.5 Summary 34
    Appendix 2-A: Derivation of T_k 34
    Appendix 2-B: Derivation of $P_k^{C_1,1}$ 35

Appendix 2-C: Derivation of T_k^1 36

3 Sensing Node Allocation in a Cognitive Radio Network with
   Cooperative Sensing **37**
   3.1   Multi-User Multi-Channel System Model 38
   3.2   Adaptive Collaboration Sensing Scheme 41
         3.2.1   Basic Idea 41
         3.2.2   Sequential Probability Ratio Test 42
         3.2.3   Optimal Sensing Node Allocation 43
   3.3   Performance Evaluation and Analysis 47
   3.4   Summary 52
   Appendix 3-A: Proof of $L(t_{ik})$ 52
   Appendix 3-B: Derivation of $P\left(\sum_{k=1}^{U_i} T_{ik} < \ln \eta_0 - \xi_i \right)$ 53

4 Transmission Power Allocation in a Cognitive Radio Network **57**
   4.1   Cognitive Radio-Assisted Cooperation Framework 58
   4.2   Optimal Transmission Power Allocation 64
         4.2.1   Cross-Layer Optimization 64
         4.2.2   Power Constraint Elimination 65
         4.2.3   Throughput Maximization 66
   4.3   Performance Analysis and Evaluation 68
         4.3.1   Simulation Scenario 68
         4.3.2   Performance Comparisons 69
         4.3.3   Impact of the Cell Population 70
         4.3.4   Impact of the Primary User Traffic Load 70
   4.4   Summary 73

5 White Space Allocation in a Cognitive Radio-Based High-Speed Vehicle
   Network **75**
   5.1   A Cognitive Radio-Based High-Speed Vehicle Network 76
         5.1.1   System Model 76
         5.1.2   Path Loss Model 80
         5.1.3   Available Channel List 82
         5.1.4   Spectrum Sharing List 83
   5.2   Maximization of Utilized White Space 83
         5.2.1   Separation Computing 85
         5.2.2   Branch and Bound Search Method 87
         5.2.3   Single-Channel Method with Low Complexity 89
         5.2.4   Linear Programming Method with Low Complexity 90
   5.3   Performance Analysis and Evaluation 92
   5.4   Summary 98

6 **Sensing Channel Allocation in a Cognitive Radio Network for a Smart Grid** **99**
   6.1 Electricity Load Shaping Framework 100
       6.1.1 Smart Grid Model 100
       6.1.2 The Cognitive Radio Network Model 104
   6.2 Sensing Channel Allocation and Load Shaping Strategies 106
       6.2.1 Sensing Channel Allocation Strategies 106
       6.2.2 Load Shaping Strategy 107
   6.3 Performance Analysis and Evaluation 109
       6.3.1 Performance of Sensing Channel Allocation 109
       6.3.2 Performance of Electricity Load Shaping 111
   6.4 Summary 116

References **117**

Index **127**

Preface

Resource allocation is an important issue in wireless communication networks. In recent decades, cognitive radio technology and cognitive radio-based networks have obtained more and more attention and have been well studied to improve spectrum utilization and to overcome the problem of spectrum scarcity in future wireless communication systems. Many new challenges on resource allocation appear in cognitive radio-based networks. In this book, we focus on effective solutions to resource allocation in several important cognitive radio-based networks, including a cognitive radio-based opportunistic spectrum access network, a cognitive radio-based centralized network, a cognitive radio-based cellular network, a cognitive radio-based high-speed vehicle network, and a cognitive radio-based smart grid.

In a cognitive radio-based opportunistic spectrum access network, secondary users hope to utilize the spectrum hole for their communications. To maximize the throughput, the secondary user wishes to access the licensed spectrum when the spectrum is detected as idle. However, to protect the primary user, it is important to prevent the secondary user from accessing the spectrum even if the spectrum is detected as idle; that is to say, there is a trade-off between throughput maximization and primary user protection. We will introduce the probabilistic slot allocation scheme to effectively allocate the transmission slots to the secondary user in order to maximize its throughput when the collision probability perceived by the primary user is constrained under a required threshold.

There are many centralized networks in practice, such as cellular networks and wireless local area networks. In the cognitive radio-based centralized network, collaborative spectrum sensing has also been well studied because the degradation of sensing performance caused by multi-path fading and shadowing can be effectively overcome via collaborative sensing. Moreover, the sensing time can also be reduced via collaborative sensing since more sensing data can be obtained simultaneously. It is challenging to design an algorithm for the fast discovery of available channels. When the number of sensing nodes is large, some nodes may be wasted if all the sensing nodes collaboratively sense one channel, since the observations from only a part of the sensing nodes are needed to determine the channel state with the predefined sensing performance guarantee. We expect to optimally allocate the appropriate number of secondary users to sense the licensed channels for fast discovery of available channels in this book.

Cooperative networking has received a significant amount of attention as an emerging network design strategy since future cellular networks are eager for higher capacity and larger coverage due to tremendously growing end-user demands and the amount of wireless terminals. A cooperative relay-aided cellular network, as a more advanced system, uses relay stations to increase capacity and coverage, which provides a better quality of service for cellular users especially at the cell edge. Moreover, cooperative relaying can also reduce overall costs compared with traditional non-relay approaches. In this book, we introduce a novel framework of cognitive radio-assisted cooperation for downlink transmissions in orthogonal frequency-division multiple access-based cellular networks and schemes to allocate transmission modes, relay stations, and transmission power/sub-channels to secondary users for throughput improvement.

It is well known that high-speed vehicles, such as high-speed trains, play an increasingly important role in people's lives, as they provide a relatively stable and spacious environment for long distance travelers. As a result, there is a strong demand for broadband wireless communications for high-speed vehicles to provide information access and onboard entertainment services for passengers. However, one design challenge is to identify sufficient spectrum resources to support broadband wireless communications in high-speed vehicles. In this book, we will introduce the cognitive radio-based high-speed vehicle network and schemes to maximize the utilized TV white space via effective allocation of white space resources to secondary users.

A smart grid is characterized by a two-way flow of electricity and information, which incorporates the benefits of distributed computing and communications to deliver real-time information and enable the near-instantaneous balance of supply and demand at the device level. Hence, real-time communications will play an important role in a smart grid. As a main candidate of communication technology, wireless communications has obtained more and more attention in a smart grid due to a number of advantages, such as wide area coverage, cost-effectiveness, quick deployment, and so forth. In a cognitive radio-based smart grid, all consumers can communicate simultaneously with utilities, which requires a high-quality spectrum sensing scheme. In this book, we introduce effective sensing channel allocation strategies for acquiring enough available spectrum, and present an analysis on the effect of communications on the performance of a novel electricity load shaping framework.

All comments and suggestions for improvements to this book are welcome.

Tao Jiang
Zhiqiang Wang
Yang Cao

Acknowledgments

The authors would like to acknowledge the support of the National Natural Science Foundation of China (NSFC), the Ministry of Science and Technology of the People's Republic of China, the Ministry of Education of the People's Republic of China, and the People's Government of Hubei Province. In particular, the majority of the work in this book is sponsored by the NSFC for Distinguished Young Scholars under Grant 61325004; the NSFC under Grants 61172052 and 60872008; the Joint Specialized Research Fund for the Doctoral Program of Higher Education (SRFDP) and Research Grants Council Earmarked Research Grants (RGC ERG) under Grant 20130142140002; the Major State Basic Research Development Program of China (973 Program) under Grant 2013CB329006; the National and Major Project under Grant 2012ZX03003004; the National High Technology Development 863 Program of China under Grants 2014AA01A704 and 2009AA011803; and the Program for New Century Excellent Talents at the University of China under Grant NCET-08-0217.

Moreover, the authors are grateful to Prof. Daiming Qu for his significant contributions on the ideas for resource allocation strategies in cognitive radio-based networks. Prof. Qu is a never-ending source of new ideas and aspects, and talking with him is always inspiring. He generously shares his broad research knowledge and experience and gives helpful suggestions. Graduate students Mr. Liang Yu, Mr. Da Chen, Mr. Chunxing Ni, Mr. Lei Zhang, Mr. Yang Zhou, and Mr. Mingjie Feng have provided a solid ground for us to stand on. Not only have we had the opportunity to work with them often on the research, but they have also provided the day-to-day discussions, both concerning research questions and other practical matters that arise when arriving at a new position. Not the least important are the travels and fun we have been through together. We are sincerely thankful to them for their great contributions to this book. Finally, we owe a great deal to our families and friends—it would have been impossible to maintain our spirit and work habits without their continuous love and support.

About the Authors

Tao Jiang (taojiang@mail.hust.edu.cn) is currently a chair professor at the School of Electronics Information and Communications, Huazhong University of Science and Technology in Wuhan, People's Republic of China (PRC). He received B.S. and M.S. degrees in applied geophysics from China University of Geosciences (Wuhan, PRC) in 1997 and 2000, respectively; and a Ph.D. in information and communications engineering from Huazhong University of Science and Technology in April 2004. For about three years, he worked in various universities, such as Brunel University (London) and the University of Michigan–Dearborn, respectively. He has authored or co-authored over 160 technical papers in major journals and conferences and six books/chapters in the areas of communications and networks. He has served or is serving as a symposium technical program committee member for some major IEEE conferences, including INFOCOM, GLOBECOM, ICC, and so forth. Jiang was invited to serve as TPC Symposium Chair for the IEEE GLOBECOM 2013 and IEEE WCNC 2013 symposiums. He has served or is serving as an associate editor for technical journals in communications, some of which include *IEEE Communications Surveys and Tutorials, IEEE Transactions on Vehicular Technology, IEEE Internet of Things Journal*, and so forth. He is a recipient of the NSFC for Distinguished Young Scholars Award in the PRC. Jiang is also a senior member of IEEE.

Zhiqiang Wang (wangzqwy@gmail.com) currently works at State Grid Shaanxi Electric Power Company Telematics. He received a B.S. from Xian Jiaotong University (Xian, PRC) in 2006, and M.S. and Ph.D. degrees from Huazhong University of Science and Technology in 2009 and 2012, respectively. Wang's current research interests include the areas of energy management and smart grid communications.

Yang Cao (ycao@hust.edu.cn) is currently an assistant professor at the School of Electronics Information and Communications, Huazhong University of Science and Technology. He received Ph.D. and B.S. degrees in information and communications engineering from Huazhong University of Science and Technology in 2014 and 2009, respectively. His research interests include resource allocation for cellular device-to-device communications and smart grids.

Chapter 1

Introduction

1.1 Cognitive Radio-Based Networks

The current communication systems are characterized by a static spectrum allocation policy according to the spectrum allocation bodies around the world. As shown
in Figure 1.1, a considerable portion of spectra have been assigned to services such as
TV/broadcasting, radio, navigation, and so forth, and quite a few spectra (labeled as
white in Figure 1.1) have not been assigned, which results in the scarcity of spectrum
resources. Hence, few spectrum resources are currently available for future wireless
applications. For example, in China, authoritative predictions indicate that the spectrum bandwidth requirement for 4G communications is 1360~1600 MHz. However,
according to the decisions of a world radiocommunications conference in 2007, only
about 448 MHz bandwidth is available for 4G communications [2].

Fortunately, spectrum surveys have shown that many parts of the assigned radio
spectrum are under-utilized [3,4] as seen in Figure 1.2, where the signal strength distribution over a large portion of the wireless spectrum is shown. In November 2002,
the U.S. Federal Communications Commission (FCC) published a report prepared by
the Spectrum Policy Task Force, whose goal was to improve the way in which this
precious resource was managed in the United States [5]. The task force was made up
of a team of high-level, multidisciplinary professional FCC staff—economists, en-
gineers, and attorneys—from across the Commission's bureaus and offices. Among
the task force's major findings and recommendations, the second finding of the report
is rather revealing in the context of spectrum utilization: "In many bands, spectrum
access is a more significant problem than physical scarcity of spectrum, in large
part due to legacy command-and-control regulation that limits the ability of potential
spectrum users to obtain such access" (p. 3). The under-utilization of the electro-
magnetic spectrum leads us to think in terms of spectrum holes. A spectrum hole is a
band of frequencies assigned to a primary user, but, at a particular time and specific

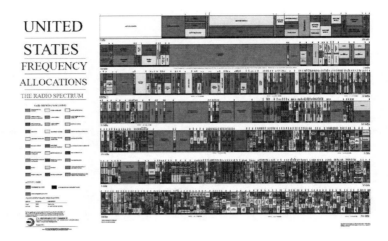

Figure 1.1: Spectrum allocation in United States. (U.S. Frequency Allocation Chart, September 10, 2012 [1].)

geographic location, the band is not being utilized by that user. Spectrum utilization can be improved significantly by making it possible for an unlicensed user to access a spectrum hole unoccupied by the licensed user at the right location and time.

Recently, dynamic spectrum access and cognitive radios have been studied and gained more and more attention due to the growing need to improve spectrum utilization [6, 7]. The term "cognitive radio" was coined by Joseph Mitola. In an article

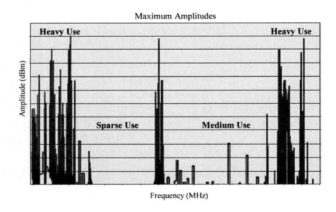

Figure 1.2: The utilization of a licensed spectrum. (I. F. Akyildiz et al., *Computer Networks*, vol. 50, no. 13, pp. 2127–2159, September 2006 [3].)

published in 1999, Mitola described how a cognitive radio can enhance the flexibility of personal wireless services through a new language called the radio knowledge representation language (RKRL) [8]. The idea of RKRL was expanded further in Mitola's own doctoral dissertation, which was presented at the Royal Institute of Technology, Sweden, in May 2000 [9]. This dissertation presents a conceptual overview of cognitive radio as an exciting multidisciplinary subject.

According to Mitola's statement above, cognitive radio signifies a radio that employs model-based reasoning to achieve a specified level of competence in radio-related domains. Specifically, cognitive radio is a goal-driven framework in which the radio autonomously observes the radio environment, infers context, assesses alternatives, generates plans, supervises multimedia services, and learns from its mistakes. This observe-think-act cycle is radically different from today's handsets that either blast out on the frequency set by the user, or blindly take instructions from the network. Cognitive radio technology thus empowers radios to observe more flexible radio etiquettes than what was possible in the past. Therefore, the cognitive radio technology requires the enhancement of current PHY and MAC protocols to adopt spectrum-agile features of allowing unlicensed users to access the spectrum hole at the right location and time.

Simon Haykin has also provided a "cognitive radio" definition [6]. As described in his statement, cognitive radio is an intelligent wireless communication system that is aware of its surrounding environment (i.e., the outside world), and uses the methodology of understanding-by-building to learn from the environment and adapt its internal states to statistical variations in the incoming radio frequency stimuli by making corresponding changes in certain operating parameters (e.g., transmit-power, carrier-frequency, and modulation strategy) in real time with two primary objectives in mind: (1) highly reliable communications whenever and wherever needed and (2) efficient utilization of the radio spectrum. Six key words stand out in this definition: *awareness*, *intelligence*, *learning*, *adaptivity*, *reliability*, and *efficiency*. Implementation of this far-reaching combination of capabilities is indeed feasible today, thanks to the spectacular advances in digital signal processing, networking, machine learning, computer software, and computer hardware.

When the users in a network can work based on cognitive radio technology, the network is recognized as cognitive radio-based network. To realize the improvement of spectrum utilization via cognitive radio technology, cognitive radio-based networks must be investigated sufficiently. Moreover, the IEEE 802.22 standard for wireless regional area networks, which was approved in July 2011, is the first practical implementation of the cognitive radio technology. The major target of this standard is to provide wireless broadband using unused VHF/UHF TV bands ranging from 54 to 862 MHz [10].

In this book, as shown in Figure 1.3, we introduce the resource allocation of several representative cognitive radio-based networks for different network scenarios, such as opportunistic spectrum access networks, centralized networks, cellular networks, high-speed vehicle networks, and smart grids. Different cognitive radio-based networks focus on different network resources, such as transmission slots, sensing nodes, transmission power, white space, and sensing channels, as shown in

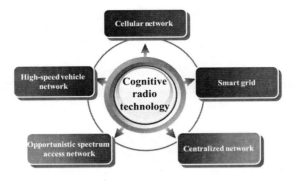

Figure 1.3: The considered CR-based networks.

Figure 1.4. These resources should be rationally allocated for reliable performance of the corresponding cognitive radio-based networks [11]. Therefore, several resource allocation schemes are introduced for different cognitive radio-based networks according to the network characteristics. In the following sections, we will introduce the challenges of different cognitive radio-based networks and related works.

1.2 Opportunistic Spectrum Access Networks

In a cognitive radio-based opportunistic spectrum access network (CR-OSAN), secondary users hope to utilize the spectrum hole for their communications. To utilize these spectrum holes with the required protection over the primary users in the same

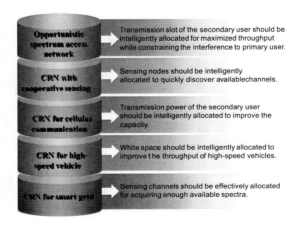

Figure 1.4: Focused resources in different CR-based networks.

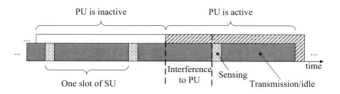

Figure 1.5: Transmission structure of a secondary user.

frequency band, secondary users have to adopt strict constraints on the resulting interference. Specifically, it is critical to maximize the secondary user's throughput under a certain primary user's collision probability constraint. Some related works studied optimal channel selection in multi-channel systems. Zhao et al. [13] proposed an optimal channel selection strategy based on the constrained Markov decision process to maximize the secondary user's throughput under a collision probability constraint over the packets of the secondary user and the primary user. Optimal sensing with access strategy was proposed by Zhao et al. to improve the secondary user's throughput, which is based on the partially observable Markov decision process. In Chen, Zhao, and Swami [15], the authors considered how to make a joint design of the spectrum sensor in the physical layer and the access strategy in the MAC layer. Optimal strategies were proposed by Chang and Liu [16] to maximize the total expected data rate of the secondary users without *a priori* knowledge of the primary user channel statistics. A method to suppress sidelobes of the orthogonal frequency division multiplexing signals was proposed, which enables the OFDM-based secondary users to utilize noncontiguous spectrum holes over multi-channels [17]. In addition to the channel selection strategies in multi-channel CR-OSAN, the throughput maximization problem has also been addressed in single-channel CR-OSAN. In Zhang and Liang [18], multiple antennas were exploited at the secondary user's transmitter to achieve an optimal trade-off between throughput maximization and interference avoidance. In Huang, Liu, and Ding [19], the structure of sensing and transmission for the secondary user is optimized to maximize its throughput under the consideration of sensing overhead. Liang et al. [20] have shown that the secondary user's throughput can be maximized by adopting an optimal sensing time. Hoang et al. [21] further considered the energy constraint at the secondary user to maximize the secondary user's throughput. A joint design between frame duration/sensing time in PHY and a random access scheme in the MAC layer was presented by Zheng et al. [22] to maximize the secondary user's throughput. The aforementioned schemes require *a priori* knowledge of the primary user traffic pattern. Huang, Liu, and Ding [24] proposed an adaptive secondary user transmission scheme to maximize the secondary user's throughput under the primary user collision probability constraint with the assumption that the primary user traffic pattern is unknown to the secondary user. However, when the parameters of the primary user traffic pattern are time-varying, the performance of this scheme is not so satisfactory.

Especially, in single-channel CR-OSAN, the instantaneous spectrum opportunity on a single channel that is licensed to a packet-based primary user should be exploited. To maximize the throughput of CR-OSAN, the secondary user is expected to transmit data if the channel is sensed as available during a certain slot. However, as shown in Figure 1.5, the interference will be introduced to the primary user when the primary user appears during a certain slot. To mitigate the interference, each transmission slot should be allocated to the secondary user with the optimal probability to maximize the throughput/spectrum utilization efficiency of the CR-OSAN with the constraint on the packet collision probability between the secondary user and the primary user.

1.3   Cognitive Radio Networks with Cooperative Sensing

Spectrum sensing for acquiring the availability of the licensed spectrum band is commonly recognized as one of the most fundamental elements in cognitive radio based networks. Spectrum sensing can be realized as a two-layer mechanism [105]. On the one hand, the PHY-layer sensing focuses on efficiently detecting the signals of primary users to identify whether the primary users are present or not. Some PHY-layer sensing methods have been studied including energy detection [43], matched filter [44], and feature detection [45]. Moreover, some sequential spectrum sensing schemes were also investigated, such as the sequential shifted chi-square test [46], which effectively reduced the average sample number compared with fixed-sample-size energy detection. Shei and Su [47] applied a sequential probability ratio test (SPRT) to control the average number of the reporting bits in cooperative sensing. On the other hand, the MAC-layer sensing, which also plays an important role in cognitive radio based centralized network, determines the channels that secondary users should sense and access in each slot for good performance, such as sensing delay, throughput of secondary users, and so on. The sensing delay results in the performance degradation of secondary users, especially in broadband communication systems since more sensing time is needed to acquire enough spectrum opportunities. Therefore, in MAC-layer sensing an important issue is how to acquire more spectrum opportunities quickly. There are many works on MAC-layer sensing. Zhao et al. [14] proposed decentralized cognitive MAC protocols and developed an analytical framework for opportunistic spectrum access based on the theory of the partially observable Markov decision process to optimize the performance of secondary users while limiting the interference perceived by primary users. Chen, Zhao, and Swami [15] jointly considered the spectrum sensor that identified spectrum opportunities, the sensing strategy that determined which channels in the spectrum to sense, and the access strategy that decided whether to access based on potentially erroneous sensing outcomes, to maximize the throughput of secondary users while limiting the probability of colliding with primary users. Hoang, Liang, and Zeng [48] adaptively scheduled the spectrum sensing periods so that negative impacts on the performance

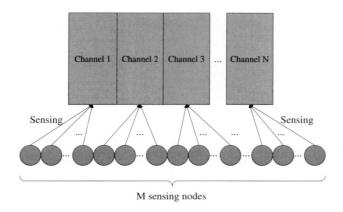

Figure 1.6: Optimal allocation of spectrum sensing nodes.

of the CR network were minimized. Kim and Shin [105] maximized the overall discovery of opportunities in licensed channels via the sensing-period optimization, and minimized the delay in locating an idle channel via the optimal channel sequencing algorithm. A sensing sequence was proposed by Kim and Shin [49] to incur a small opportunity discovery delay by considering the probability of the spectrum band being available at the sensing time, the duration of sensing on a channel, and the channel capacity. In Kim and Giannakis [50], sequential sensing algorithms were developed to maximize the average total achievable rate of a multi-channel CR system while considering the time spent on sensing. Oksanen, Lunden, and Koivunen [51] proposed a machine-learning-based multi-band spectrum sensing policy, in which they exploited the information about the sensing performance of secondary users to optimize the sensing assignments in an energy efficient manner.

Collaborative spectrum sensing has also been well studied in cognitive radio-based centralized networks in recent years since the degradation of sensing performance caused by multi-path fading and shadowing can be effectively overcome via collaborative sensing. Moreover, the sensing time can also be reduced via collaborative sensing since more sensing data can be obtained simultaneously. In multi-user multi-channel cognitive radio-based centralized networks, there are new challenges to designing algorithm for the fast discovery of available channels. When the number of sensing nodes is large, some nodes may be wasted if all sensing nodes collaboratively sense one channel, since the observations from only a part of sensing nodes are needed to determine the channel state with the predefined sensing performance guarantee. Therefore, it is important to optimally allocate the resources of sensing nodes for fast discovery of available channels, as shown in Figure 1.6.

1.4 Cognitive Radio Networks for Cellular Communications

Cooperative networking has received significant attention as an emerging network design strategy since the future cellular networks are eager for higher capacity and larger coverage due to tremendously growing end-user demands and the amount of wireless terminals. One of the ways is to deploy more base stations (BSs) as taken in traditional cellular networks. In contrast, a cooperative relay-aided cellular network, as a more advanced system, introduces the use of relay stations (RSs) to increase the capacity and coverage, which provides a better quality of service (QoS) for cellular users (CUs) especially at the cell edge [61]. Moreover, cooperative relaying can also reduce overall cost compared with the traditional non-relay approach [62]. Recently, the resource allocation in relay-aided cellular networks has been investigated. An objective function of the total average throughput with both the direct and relayed links was proposed in Nam et al. [33], in which two algorithms were proposed to improve the overall cell-throughput while minimizing the system complexity through optimal joint power and subcarrier allocation. To maximize network sum utility, a joint optimization of the tone assignment, relay strategy selection, and power allocation in each tone of orthogonal frequency division multiple access (OFDMA) cellular networks were proposed by Ng and Yu [69]. In Kim and Lee [34], a semi-distributed downlink OFDMA scheme in a single cell enhanced by some half-duplex fixed RSs was considered, in which CUs in the neighborhoods of the BS and RSs are referred to as the BS-CU and RS-CU clusters, respectively. Then the BS directly allocates some resources to the BS-CU cluster and the RS-CU cluster through the RS. The problem of sub-channel assignment and power allocation for multi-hop OFDMA networks were studied in depth by Kim, Wang, and Madihian [35]. Sundaresan and Rangarajan [36] considered the problem of scheduling users with backlogged and finite buffers over multiple sub-carriers in the relay-aided OFDMA cellular networks in order to leverage spatial reuse. A network coding assisted cooperation scheme was proposed by Xu and Li [37], in which full-duplex cooperation is adopted with the assumption of the existence of dedicated sub-channels for the RS-to-CU links. With the assumption that the sub-channels are preassigned, the authors proposed a joint optimization of relay selection and power allocation [63] in which a related convex optimization problem was formulated, which provided an extremely tight upper bound on system performance. Li et al. [38] proposed cooperative communication schemes with rateless network coding in multple-in multiple-out (MIMO)-based cellular networks. One critical observation from existing work is that the performance gain of adopting cooperative relaying in cellular networks can be minor due to the following two major reasons (we use a downlink transmission case for illustration).

■ **Bottleneck link between the base station and cellular user**: Generally, the link between the base station and the relay station (BS-to-RS link) is line-of-sight as the base station and the relay station are both placed at some height above the ground, while the link between the base station and the cellular user (BS-to-CU link) and the link between the relay station and the cellular user (RS-to-CU link) are non-line-of-sight. Therefore, the BS-to-RS link has

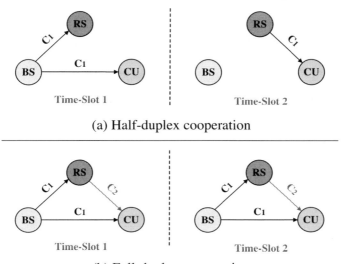

Figure 1.7: An example of motivation.

better quality than the BS-to-CU and RS-to-CU links most of the time. As a result, the performance gain of the cooperative relaying is constrained by the quality of the weaker link, that is, the compound BS/RS-to-CU link [61].

■ **Half-duplex cooperation**: Obviously, direct transmissions of BS-to-RS/CU and RS-to-CU can not be scheduled simultaneously. As depicted in Figure 1.7a, the base station and relay station are scheduled sequentially to transmit at two different time slots over the same sub-channel C_1. Such time-division multiplexing basically leads to a 50% throughput reduction of the transmission from BS to CU and may offset the potentially achievable benefit from cooperative diversity [63].

A novel way is to leverage a cognitive radio technique at the relay station to make concurrent BS-to-RS/CU and RS-to-CU transmissions possible (as depicted in Figure 1.7b), and in turn overall downlink network throughput can be improved in contrast to traditional cooperative relaying. The target system, based on the orthogonal frequency-division multiple access, is a cooperative cellular network assisted by a cognitive radio technique. Recently, some work has been done to combine the cognitive radio and the relay-aided cellular network to implement network goals in a coordinated way. In Jia, Zhang, and Zhang [74], a joint relay selection and spectrum allocation was formulated and solved in the context of cooperative relaying in cognitive radio cellular networks. Kim et al. [39] studied RSs to exploit white spaces in an opportunistic manner, in which the authors supposed that the RSs were connected with the cellular core network, resulting in the requirement

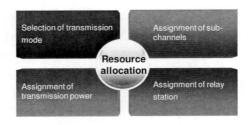

Figure 1.8: Resource allocation in the downlink of a cellular network.

of a higher infrastructure cost. A distributed resource scheduling scheme was proposed in the relay-assisted cognitive cellular network [40], where only white space channels can be utilized. A network architecture was proposed by Sachs, Mari, and Goldsmith [41], to enable spectrum sharing between a primary TV broadcast system and a secondary cellular broadband system. In this framework, a cellular base station spends part of its power to cooperate with the TV system to obtain the access right of the TV bands with favorable propagation properties. Luo et al. [42] studied cooperative diversity techniques in the cognitive radio environment by considering radio resources extended from time-frequency dimensions to space-time-frequency dimensions.

However, a centralized optimization framework is needed to maximize the network utility in the framework. Joint resource allocation that includes transmission mode selection, relay station allocation, and transmit power/sub-channel allocation is considered to cost-effectively provide services and applications, as shown in Figure 1.8.

1.5   Cognitive Radio Networks for High-Speed Vehicles

It is well known that high-speed vehicles, such as high-speed trains, play an increasingly important role in people's lives, as they provide a relatively stable and spacious environment for long distance travelers. As a result, there is a strong demand of broadband wireless communications (BWC) for high-speed vehicles to provide information access and onboard entertainment services for passengers. However, the cognitive radio network should be employed to overcome the problem of spectrum scarcity. For cognitive radio networks, primary channel occupancy modeling is very important for designing secondary user spectrum accessing and scheduling schemes. Primary users (PUs) in different primary networks usually have different activity models, such as in cellular networks [76], wireless local area networks (WLAN) [27], TV/broadcasting networks [77], and so forth. However, the existing works are suitable for static or low-speed mobile users, and do not consider situations with high-speed mobile users, in which the channel availabilities are very different due to the high velocity of the users [78, 79]. In the existing works, spectrum resource allocation in cognitive radio has also been well researched. Wang and Liu [80] studied the available channel dynamics caused by the location and traffic load of the primary users. Then, they formulated the allocation of the available dynamic channels as a list multi-coloring problem. They also proposed a distributed greedy algorithm, a distributed fair algorithm, and a distributed randomized algorithm, to obtain suboptimal solutions with low complexity. Zheng and Peng [81] developed a graph-theoretical model to characterize the spectrum access problem under a number of different optimization functions. Then, they discussed how to employ a set of heuristic-based approaches to avoid non-deterministic polynomial (NP) complexity. In Yang and Wang [82], the radio allocation is formulated as a large-scale combinatorial optimization problem, which is solved using a proposed sequential optimization scheme based on an investigation of the conditions that the optimal solution should satisfy.

Obviously, the existing works on TV white space allocation are only suitable for static or low-speed mobile users, and do not consider situations with high-speed mobile users, in which the channel availabilities are very different due to the high velocity of the users. Hence, it is important to study cognitive radio-based high-speed vehicle networks that can effectively utilize TV white space.

1.6 Cognitive Radio Networks for a Smart Grid

In the power grid, demand side management is considered to shape the power load, for more stability, this is, during a certain period of time, the electricity demand is redistributed for the balance of supply and demand. There are some load shaping methods including peak clipping, valley filling, strategic conservation, and load shifting [91]. Usually, these methods can be divided into two categories. One is direct load control implemented by utilities, such as peak clipping. The other is indirect load control by motivating customers to use more electricity when valley load appears and less electricity when peak load appears, such as load shifting. Dynamic pricing is considered as one of the effective motivation strategies for load shaping, especially in a smart grid, which has attracted more and more attention due to the rising cost of energy and the urgent need to reduce global carbon emissions [92]. As presented by the U.S. Department of Energy [93], a smart grid will be characterized by a two-way flow of electricity and information, and will introduce the benefits of distributed computing and communications into the grid so as to deliver real-time information and enable the near-instantaneous balance of supply and demand at the device level. Hence, real-time communications will play an important role in a smart grid.

As a main candidate of communication technology, wireless communication has obtained more and more attention in a smart grid due to many advantages, such as wide area coverage, cost-effectiveness, quick deployment, and so forth [94–96]. However, some existing wireless communication technology suffer from several drawbacks when used in a smart grid [96]. First, the popular wireless communication standard 802.11, which employs the Industrial Scientific Medical (ISM) unlicensed spectrum bands, is not suitable for the smart grid, since the ISM bands are heavily utilized in urban areas while not well suitable for the distance requirements in rural areas. Second, as the most successful commercial communication network, cellular network is also not suitable for the smart grid, since the extra expense associated with licensed bands is needed in cellular network. Moreover, there is considerable competition for this bandwidth in urban areas and limited availability in rural areas. Third, the use of proprietary mesh network technology reduces inter-operability and impedes meter diversity [96]. Hence, new wireless communications service suitable for smart grid should be investigated. Unfortunately, spectrum scarcity is a big challenge for new wireless communication services, since the current wireless communication systems are characterized by a static spectrum allocation policy according to the spectrum allocation bodies around the world, and few spectrum resources are currently available for new wireless applications. However, a survey [3] shows that current spectrum utilization is very inefficient. Recently, dynamic spectrum access and cognitive radios have been proposed and have gained more and more attention due to the growing need to improve spectrum utilization and solve the scarcity problem of spectrum resources [6, 7]. Some literature has tried to combine cognitive radio technology with a smart grid [96–102]. Fatemieh, Chandra, and Gunter [96] investigated the use of white space in the TV spectrum for advanced meter infrastructure (AMI) communications in a smart grid, and showed its benefits in terms of bandwidth, deployment, and cost. Ghassemi, Bavarian, and Lampe [97] proposed the application of cognitive radios based on the IEEE 802.22 standard in smart grid wide area networks, and discussed the benefits of the proposed scheme including opportunistic access of TV bands, extended coverage, ease of upgradability, self-healing, and fault-tolerant design. Brew et al. [98] presented a white space communications test bed running in the Scottish Highlands and Islands, and discussed its feasibility for smart grid communications. Nagothu et al. [99] used a cloud computing data center as the central communication and optimization infrastructure supporting a cognitive radio network of AMI. Yu et al. [100] presented an unprecedented cognitive radio-based communications architecture for the smart grid, which was decomposed into three subareas: cognitive home area network, cognitive neighborhood area network, and cognitive wide area network. In Vo et al. [101], cognitive radio functions based on different multi-objective genetic algorithms for smart meters were proposed for finding the trade-off between power efficiency and spectrum efficiency in different operating environments. A dynamic spectrum allocation scheme was proposed by Gong and Li [102] for the power load prediction.

Note that the cognitive radio-based network requires the enhancement of current PHY and MAC protocols to adopt spectrum-agile features, which are to allow secondary users to access the licensed spectrum band when the primary users are absent. In a smart grid, all consumers communicate with utilities nearly simultaneously, which requires a high quality of spectrum sensing scheme when cognitive radio technology is employed in a smart grid. Specifically, effective sensing channel allocation for acquiring enough available spectrum is commonly recognized as one of the most fundamental issues in a cognitive radio-based smart grid.

1.7   Content and Organization

We will introduce the effective resource allocation strategy in cognitive radio-based networks. The rest of the book is organized as follows.

In Chapter 2, a single secondary user and a single licensed channel are considered, and we optimally allocate transmission slots for the secondary user in the licensed channel to maximize its throughput when the collision probability perceived by the primary user is constrained under the required threshold. To achieve this goal, we introduce a novel probabilistic slot allocation scheme for opportunistic spectrum access based cognitive radio system. Moreover, we study the maximum achievable secondary user throughput under perfect and imperfect sensing situations, respectively. The distribution of the primary user idle period is also analyzed.

In Chapter 3, multiple secondary users and multiple licensed channels are considered, and we optimally allocate the appropriate number of secondary users to sense the licensed channels for fast available channel discovery. To acquire more spectrum opportunities in a limited sensing time, we construct a novel MAC-layer sensing framework for efficient acquisition of spectrum opportunities. Specifically, we employ the sequential probability ratio test and develop a new collaboration sensing scheme for multi-users to collaborate during multi-slots, in which we effectively utilize the resources of secondary users to sense the channels for efficient acquisition of spectrum opportunities.

In Chapter 4, the resource allocation in the downlink of the cognitive radio network for cellular communication is considered. We allocate transmission modes, relay stations, and transmission power/sub-channels to secondary users for throughput improvement. We introduce a novel framework of cognitive radio-assisted cooperation for downlink transmissions in orthogonal frequency-division multiple access-based cellular communication systems, in which relay stations are deployed in each cell and have spectrum sensing capability. One of the promising novelties is that our scheme considers joint resource allocation, which includes transmission mode selection, relay station allocation, and transmit power/sub-channel allocation to provide cost-effective services and applications.

It is well known that broadband wireless communications (BWC) are necessary for high-speed vehicles since passengers need many broadband wireless multimedia services. However, one design challenge is to identify sufficient spectrum resources to support BWC in high-speed vehicles. In Chapter 5, cognitive radio is considered as a promising technology to solve the scarcity problem of spectrum resources. In the cognitive radio-based high-speed vehicle network, we optimally allocate the white space to high-speed vehicles for maximum utilization of white space. We introduce a white space resource allocation framework, where high-speed vehicles can effectively utilize the TV white space.

In Chapter 6, the sensing channel allocation in a smart grid is introduced. In the cognitive radio-based smart grid, the communication scenario has some new characteristics. We introduce two sensing channel allocation strategies for the communication between utilities and the electricity consumer to support a novel effective electricity load shaping scheme. Based on the effective communication, the utilities obtain the load information of all consumers via cognitive radio-based wireless communications. Then, according to the load information, the price of electricity is determined by a dynamic pricing scheme to encourage the consumers to rationally use energy storage for achieving load shaping.

Chapter 2

Transmission Slot Allocation in an Opportunistic Spectrum Access Network

In a cognitive radio-based opportunistic spectrum access network (CR-OSAN), secondary users hope to utilize the spectrum hole for their communications. To utilize these spectrum holes with the required protection over primary users in the same frequency band, secondary users have to adopt strict constraints on the resulting interference. For example, for the packet-based primary user, CR-OSAN usually constrains the probability of collision with the primary user's packet to be under a predefined threshold. Therefore, it is critical to know how to maximize the secondary user's throughput under a certain primary user's collision probability constraint.

In a single-channel CR-OSAN, the instantaneous spectrum opportunities on a single channel that is licensed to a packet-based primary user should be exploited. To maximize the throughput of CR-OSAN, the secondary user is expected to transmit data if the channel is sensed as available during a certain slot. However, the interference will be introduced to the primary user when the primary user appears during a certain slot. To mitigate the interference, we should allocate each transmission slot to the secondary user with the optimal probability to maximize the throughput/spectrum utilization efficiency of the CR-OSAN with the constraint on the packet collision probability between the secondary user and the primary user.

In this chapter, we introduce an effective transmission slot allocation scheme that exploits instantaneous spectrum opportunities on a single channel that is licensed to a packet-based PU [23]. The scheme schedules the transmission probabilities of the secondary user (SU) during an idle period of the primary user (PU), which spans over some SU slots. Our aim is to maximize the throughput/spectrum utilization efficiency of the CR-OSAN with the constraint on the packet collision probability between SU and PU. When the spectrum sensing at the secondary user is perfect and the primary user's traffic pattern is known to the secondary user, we introduce a probabilistic slot allocation scheme for transmission slot allocation to maximize the secondary user's throughput under the primary user collision probability constraint. We also quantify the impact of sensing errors on the secondary user performance with the probabilistic slot allocation scheme when the spectrum sensing at the secondary user is imperfect. When the primary user's traffic pattern is unknown to the secondary user and is time-varying, along with errors in spectrum sensing, we develop a method to predict the future primary user states by learning past periodic sensing results based on the hidden Markov model, and use these predictions in the probabilistic slot allocation scheme.

2.1   Single-User Single-Channel System Model

As depicted in Figure 2.1, the SU exploits spectrum holes while limiting the interference perceived by the PU who works on the same channel in a CR-OSAN. The PU has the priority to access the channel and it is not responsible for any sort of notification to the SU for its transmission. We assume that the locations of the PU transmitters/receivers and the SU transmitters/receivers disperse in the same region with a limited area, that is, the delay for packet transmission is negligible. Suppose that the PU traffic pattern follows an ON/OFF model (also adopted in Huang, Liu, and Ding [24]), in which its state becomes busy ("ON," packet transmission) or idle ("OFF," idle period/spectrum hole) alternatively. Thus, the SU should target a temporal utilization of the spectrum hole and vacate the channel as quickly as possible when the PU state turns from idle to busy.

Suppose that the SU is equipped with a half-duplex transceiver, that is, during data transmission, the SU is no longer capable of sensing the channel. To schedule sensing and transmission periodically, the SU employs a slotted communication protocol [13] as depicted in Figure 2.2, in which L denotes the time duration for one SU slot. At the beginning of each slot, the SU turns off the transmitter and senses the channel during the sensing sub-slot that lasts for L_s; when the duration of the sensing sub-slot is over, the SU transmits data during the following data sub-slot that lasts for L_d, $L = L_s + L_d$. Generally, $L_s \ll L_d$. To protect the PU packets, the SU is allowed to access the channel only when the current sensing result shows that the channel is not occupied by the PU.

For a CR-OSAN, a strict interference constraint is usually imposed on the SU to ensure that the interference perceived by the PU lies below a predefined threshold. Due to the fact that the SU cannot sense the channel during data transmission, a sensing result that shows the PU is idle during the current sensing sub-slot does not mean that the idle state of the PU will remain during the following data sub-slot.

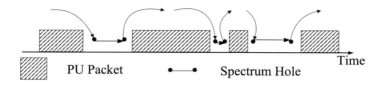

Figure 2.1: Spectrum holes in a CR-OSAN.

As a result, collision occurs if the PU accesses the channel when the SU is transmitting during the data sub-slot, consequently resulting in the packets loss for both PU and SU. If the spectrum sensing at the SU is perfect, that is, the probability of sensing errors approaches zero, collisions only occur at the head of a PU packet, which is denoted as a type-I collision; otherwise, collisions may happen during the PU transmission due to missed detection, which is denoted as a type-II collision, as shown in Figure 2.2. To protect the PU packets from excessive collisions, we define the average ratio of collisions in all primary packets during a certain time duration U as a measure of the interference caused by the SU from the PU's perspective (a similar definition of interference metric has also been adopted in Huang, Liu, and Ding [24]), denoting the average packet collision ratio (APCR) as R_C, that is,

$$R_C = \lim_{U \to +\infty} \frac{N_C}{N_P}, \tag{2.1}$$

where N_P is the number of the PU packets during U, and N_C is the number of collisions during U. In this chapter, we aim to constrain R_C to be below a predefined threshold R_{TH}. In other words, the PU can tolerate the interference and can safely transmit data when $R_C \leq R_{TH}$.

Moreover, we are concerned with the SU throughput, which depends on the utilization efficiency of the spectrum hole. In this chapter, we employ spectrum hole utilization efficiency as a measure of the throughput for the CR-OSAN (a similar definition of the SU throughput has also been adopted in Huang, Liu, and Ding [24]).

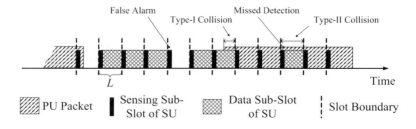

Figure 2.2: The structure of the SU transmission.

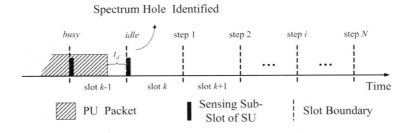

Figure 2.3: Spectrum hole identification.

Therefore, the normalized throughput of the SU is defined as

$$T = \lim_{U \to +\infty} \frac{N_S}{N_{IDLE}}, \tag{2.2}$$

where N_S is the total number of data sub-slots that are successfully utilized by the SU in the PU idle period during U, and N_{IDLE} is the total number of available data sub-slots in the PU idle period during U.

2.2 Probabilistic Slot Allocation Scheme

In this section, we introduce a probabilistic slot allocation (PSA) scheme for the CR-OSAN, in which the SU optimally schedules the transmission probabilities for the data sub-slots during the PU's idle period. With a PSA scheme, the SU's throughput is maximized while the APCR perceived by the PU lies below a preset threshold.

Obviously, spectrum sensing results are required for the SU to identify a spectrum hole, in which the sensing result is binary for every sensing sub-slot, that is, either busy or idle. To simplify the analysis, we suppose the sensing results are perfect in this section.[1] The sensing result $X(k)$ for the k-th (k is a positive integer) sensing sub-slot is written as

$$X(k) = \begin{cases} 1, & \text{if busy} \\ 0, & \text{if idle}. \end{cases} \tag{2.3}$$

As depicted in Figure 2.3, a spectrum hole is identified by the SU during the k-th sensing sub-slot when the current sensing result $X(k) = 0$ and the previous sensing result $X(k-1) = 1$ ($k > 1$). Then, the SU predicts the PU channel occupancy states in the following N sensing sub-slots, where $N \in \mathbf{N}^*$ is the number of prediction steps and ideally $N \to +\infty$. Based on the prediction, the SU schedules its transmission probabilities in the following N data sub-slots. If the k-th to the $(k+i)$-th sensing sub-slot are all idle, the SU transmits its packet in the $(k+i)$-th data sub-slot with the probability $P_k^T(i)$, $i = 0, 1, ..., N-1$. Otherwise, the SU stops its transmission and

---
[1] The effect of sensing errors will be quantified in Section 2.4.1.

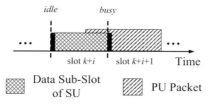

Figure 2.4: Collision between the PU and the SU.

waits for the next spectrum hole. If the PU idle period is longer than N slots, the SU transmission probabilities on the data sub-slots after the first N data sub-slots are set to zero.

Next, we present how to obtain the optimal transmission probabilities to maximize SU throughput with the constraint of the APCR.

First, we derive the expected collision probability P_k^C of the coming PU packet. As shown in Figure 2.4, the SU transmits during the $(k+i)$-th data sub-slot, if $X(k+i+1) = 1$, it is reasonable to consider that the PU accesses the channel again during the $(k+i)$-th data sub-slot and a collision occurs; otherwise, the SU transmission during the $(k+i)$-th data sub-slot is successful.

With the transmission probabilities $P_k^T(i)$, we have

$$P_k^C = \sum_{i=0}^{N-1} B_k(i) \cdot P_k^T(i), \tag{2.4}$$

where $B_k(i)$ is the probability that the PU starts to access the channel during the $(k+i)$-th data sub-slot, $i = 0, 1, ..., N-1$. Due to the relatively short duration of the sensing sub-slot, we ignore the situation that the PU state changes during a sensing sub-slot.

To limit the interference to the PU, R_{TH} is preset as a constraint on the APCR. Obviously, when R_{TH} is the threshold of P_k^C, that is, $P_k^C \leq R_{TH}$ for every PU packet, then $R_C \leq R_{TH}$ is satisfied.

Second, we derive the expected normalized throughput T_k of the SU in the idle period starting from the k-th sensing sub-slot. During the $(k+i)$-th data sub-slot, if the SU transmits and no collision occurs, the data sub-slot has been successfully utilized; otherwise, the SU transmission has failed. $I_k(i)$ is the probability that the PU will not access the channel from the k-th to the $(k+i)$-th data sub-slot, $i = 0, 1, ..., N-1$. Thus, we have

$$T_k = \frac{\sum_{i=0}^{N-1} I_k(i) \cdot P_k^T(i)}{\sum_{i=0}^{N-1} I_k(i)}, \tag{2.5}$$

where $\sum_{i=0}^{N-1} I_k(i)$ is the expectation of the PU idle period length (Appendix 2-A shows the derivation of T_k in detail). Obviously, if we maximize T_k in every idle period, the normalized SU throughput T is also maximized.

Then, the optimal transmission probabilities that maximize the SU throughput is formulated as follows

$$\begin{aligned} \text{Max.} \quad & T_k \\ \text{s.t.} \quad & P_k^C \le R_{\text{TH}} \\ & 0 \le P_k^T(i) \le 1, i = 0, ..., N-1. \end{aligned} \tag{2.6}$$

It is obvious that (2.6) is a typical linear programming problem [28]. Combining (2.4), (2.5), and (2.6), we first derive $B_k(i)$ and $I_k(i)$ for the solution of (2.6). Suppose that the PU idle period follows a specific distribution with the probability density function (PDF) $f(t)$, where t is the time that elapses from the beginning ($t = 0$) of this idle period. Let $F(t)$ denote the cumulative distribution function (CDF) of the PU idle period, and the complementary CDF (CCDF) is $\overline{F}(t) = 1 - F(t)$. We then have $I_k(i)$ under condition t_d as

$$I_k(i)\big|_{t_d} = \frac{\Pr\{t > [t_d + (i+1)L]\}}{\Pr(t > t_d)} = \frac{1 - F[t_d + (i+1)L]}{1 - F(t_d)}$$

$$= \frac{\overline{F}[t_d + (i+1)L]}{\overline{F}(t_d)}, \quad 0 \le i \le N-1, \tag{2.7}$$

where t_d denotes the delay of the spectrum hole identification, that is, the time interval between SU identifying an idle period and the actual beginning of this idle period, as shown in Figure 2.3. Suppose that t_d is uniformly distributed over the interval $(0, L)$, then,

$$I_k(i) = E_{t_d}[I_k(i)\big|_{t_d}]. \tag{2.8}$$

Therefore,

$$B_k(i) = \begin{cases} 1 - I_k(i), & i = 0, \\ I_k(i-1) - I_k(i), & 0 < i \le N-1. \end{cases} \tag{2.9}$$

Obviously,

$$\lim_{N \to \infty} \sum_{i=0}^{N-1} B_k(i) = 1. \tag{2.10}$$

When N is large enough (but finite), $\sum_{i=0}^{N-1} B_k(i) \approx 1$ and $\sum_{i=N}^{+\infty} B_k(i) \approx 0$.

Assuming that the SU has *a priori* knowledge of $f(t)$, $B_k(i)$ and $I_k(i)$ can be directly computed by (2.8) and (2.9), then optimal $P_k^T(i)$ is obtained from (2.6). In Section 2.4.2, we introduce how to predict $B_k(i)$ and $I_k(i)$ without *a priori* knowledge of $f(t)$.

In summary, the PSA scheme contains three operating stages. The first stage is spectrum hole identification. Once a spectrum hole is identified, the PSA scheme immediately goes to the second stage, in which the SU obtains $B_k(i)$ and $I_k(i)$. In the final stage, the SU optimally schedules its transmission probabilities during the following N data sub-slots based on $B_k(i)$ and $I_k(i)$.

2.3 Optimal Probabilistic Slot Allocation

In this section, we study the baseline performance of the SU throughput without knowing the distribution of the PU idle period. When the distribution is *a priori* knowledge to the SU, the achievable SU throughput with different distributions of the PU idle period for a PSA scheme is presented.

2.3.1 Baseline Performance

In this section, we employ a baseline scheme to solve (2.6), which does not require *a priori* knowledge of the PU idle period distribution. For a baseline scheme, the transmission probabilities $P_k^{\mathrm{T}}(i) = R_{\mathrm{TH}}$ for all data sub-slots during a PU idle period. Thus, according to (2.4) and (2.10), the resulting APCR is

$$P_k^{\mathrm{C}} = \lim_{N \to \infty} \sum_{i=0}^{N-1} B_k(i) \cdot R_{\mathrm{TH}} = R_{\mathrm{TH}}. \tag{2.11}$$

Therefore, the normalized throughput of the SU during the idle period starting from the k-th sensing sub-slot is

$$T_k = \lim_{N \to \infty} \frac{\sum_{i=0}^{N-1} I_k(i) \cdot R_{\mathrm{TH}}}{\sum_{i=0}^{N-1} I_k(i)} = R_{\mathrm{TH}}. \tag{2.12}$$

Thus, the baseline performance T_k equals to the APCR threshold R_{TH}.

In the following two sections, we study the maximum achievable throughput of the SU with a PSA scheme when we assume different distributions of the PU idle period.

2.3.2 Exponential Distribution

The Poisson arrival traffic model has been widely used in the literature [13–15, 19]. In this model, the idle period follows an exponential distribution and its PDF is

$$f(t) = \lambda e^{-\lambda t}, \ t \ge 0, \tag{2.13}$$

where λ is the rate parameter. According to (2.8) and (2.9), we have

$$\begin{aligned}
I_k(i)\big|_{t_d} &= \frac{\overline{F}[t_d + (i+1)L]}{\overline{F}(t_d)} = \frac{e^{-\lambda[t_d + (i+1)L]}}{e^{-\lambda t_d}} \\
&= e^{-\lambda(i+1)L}, \ i = 0, 1, ..., N-1
\end{aligned} \tag{2.14}$$

then,

$$I_k(i) = E_{t_d}[e^{-\lambda(i+1)L}] = e^{-\lambda(i+1)L}, \tag{2.15}$$

$$B_k(i) = (e^{\lambda} - 1) \cdot e^{-\lambda(i+1)L}. \tag{2.16}$$

Therefore, we have

$$I_k(i) = \eta B_k(i), \tag{2.17}$$

where $\eta = (e^\lambda - 1)^{-1}$ is a constant.

From (2.5), (2.10), and (2.17), the normalized throughput is expressed as

$$T_k = \frac{\lim\limits_{N\to\infty} \sum\limits_{i=0}^{N-1} \eta B_k(i) \cdot P_k^T(i)}{\lim\limits_{N\to\infty} \sum\limits_{i=0}^{N-1} \eta B_k(i)} = \frac{P_k^C}{\lim\limits_{N\to\infty} \sum\limits_{i=0}^{N-1} B_k(i)} \leq R_{\text{TH}}. \tag{2.18}$$

Obviously, the maximum normalized throughput is

$$T_k^{\max} = R_{\text{TH}}. \tag{2.19}$$

Therefore, for the Poisson arrival traffic model, the maximum normalized throughput is the same as that of the baseline scheme, which equals to R_{TH}.

2.3.3 Hyper-Erlang Distribution

In wireless networks, the traffic flows are commonly considered to be self-similar. The traditional Poisson arrival traffic model likely becomes invalid to model such kind of traffic. The hyper-Erlang model is a natural choice for self-similar traffic modeling in communications networks with integrated services, which is a better fit than exponential distribution for wireless networks [25–27], especially for the modeling of white space for an 802.11b-based wireless local area network [27].

Suppose the hyper-Erlang-k-2 distribution contains two weighted Erlang-k components for the duration of the PU idle period, and its PDF is

$$f(t) = \alpha_1 \frac{(k_1\lambda_1)^{k_1} t^{k_1-1}}{(k_1-1)!} e^{-k_1\lambda_1 t} + \alpha_2 \frac{(k_2\lambda_2)^{k_2} t^{k_2-1}}{(k_2-1)!} e^{-k_2\lambda_2 t}, \tag{2.20}$$
$$t \geq 0,$$

where λ_1 and λ_2 are rate parameters, k_1 and k_2 are shape parameters, and α_1 and α_2 are weights with $\alpha_1 + \alpha_2 = 1$. Therefore, we have

$$\overline{F}(t) = \alpha_1 \sum_{i=0}^{k_1-1} \frac{(k_1\lambda_1 t)^i}{i!} e^{-k_1\lambda_1 t} + \alpha_2 \sum_{i=0}^{k_2-1} \frac{(k_2\lambda_2 t)^i}{i!} e^{-k_2\lambda_2 t}. \tag{2.21}$$

Some distribution parameters are listed in Table 2.1, the APCR threshold $R_{\text{TH}} = 0.1$ and the prediction steps $N = 40$. We then have $T_k = 0.1984$ with the optimal solution to (2.6), while $T_k = 0.1$ for the baseline scheme. We find that the maximum achievable throughput of the SU that adopts a PSA scheme is beyond that adopting the baseline scheme for nearly 100%.

In summary, we conclude that:

■ The distribution of the PU idle period has a great effect on the maximum achievable throughput of the SU;

Table 2.1   Hyper-Erlang Distribution Parameters

Parameter	k_1	k_2	λ_1 [slot^{-1}]	λ_2 [slot^{-1}]	α_1	α_2
Value	2	2	1/6	1/10	0.5	0.5

- The baseline scheme achieves APCR constraining and a normalized throughput equals to R_{TH} without *a priori* knowledge of the PU idle period distribution;

- For the exponential distribution of the PU idle period, maximum achievable throughput is the same as that of the baseline scheme;

- For the hyper-Erlang distribution of the PU idle period, the throughput can be significantly improved with the PSA scheme compared with the baseline scheme.

2.4   Performance Analysis and Evaluation

2.4.1   *Impact of Sensing Errors*

Generally, the spectrum sensing of the SU is not perfect in a CR-OSAN. Therefore, it is indeed necessary to consider the impact of sensing errors to the PSA scheme.

Let $\{P_f, P_m\}$ denote the sensing errors, where P_f is the probability of a false alarm (declare the idle channel as occupied by the PU), while P_m is the probability of missed detection (declare the channel that is occupied by the PU as idle). It is reasonable to suppose the events of sensing errors are independent across time.

Now, we discuss the impact of sensing errors on the APCR perceived by the PU. Let P_k^{C1} and P_k^{C2} denote the probabilities of type-I and type-II collisions of the PU packet after the idle period starting at the slot k, respectively. Thus, the expected collision probability is

$$P_k^{\text{C}} = P_k^{\text{C1}} + P_k^{\text{C2}}. \tag{2.22}$$

First, we derive the P_k^{C2}. Let N_m denote the number of missed detection events during a PU packet, and $\Pr\{N_m = n\}$ represents the probability that the missed detection events occur n times during the PU packet. Therefore, we have

$$\Pr\{N_m = n\} = E_{L_p}\left[\binom{\lceil L_p \rceil}{n} P_m^n (1 - P_m)^{\lceil L_p \rceil - n} \right], \tag{2.23}$$

where L_p is the number of SU slots that a PU packet spans over, and its expectation $E[L_p] = l_p, (l_p \in \mathbf{R})$, where $\lceil g \rceil$ denotes the nearest integer greater than or equal to g. Let $P_k^{\text{C2,n}}$ denote the type-II collision probability when missed detection events occur n times. Therefore,

$$P_k^{\text{C2}} = \sum_{n=0}^{+\infty} \Pr\{N_m = n\} \cdot P_k^{\text{C2,n}}. \tag{2.24}$$

If P_m is small enough (e.g., $P_m \leq 0.05$), $\Pr\{N_m = n\}$ is negligible when $n \geq 2$. As a result, we only need to consider the cases of $n \leq 1$. Therefore, we have

$$P_k^{C_2} \approx \Pr\{N_m = 0\} \cdot P_k^{C_2,0} + \Pr\{N_m = 1\} \cdot P_k^{C_2,1}. \tag{2.25}$$

Obviously, $P_k^{C_2,0} = 0$. For $P_k^{C_2,1}$, we consider two cases. For the first case, the missed detection occurs at the first sensing sub-slot during a PU packet, then, the transmission probability during the following data sub-slot might be low due to the fact that the SU considers the PU idle state during the sensing sub-slot as a part of the current idle period rather than the beginning of a new idle period. As a result, the contribution of this case to the $P_k^{C_2,1}$ can be neglected. For the second case, the missed detection happens at other sensing sub-slots during a PU packet, where the SU transmits with probability $P_k^T(0)$. Thus, the collision probability is $P_k^T(0)$. Therefore, we have

$$P_k^{C_2,1} \approx \frac{l_p - 1}{l_p} \cdot P_k^T(0). \tag{2.26}$$

When P_m and l_p are small enough, we obtain

$$\Pr\{N_m = 1\} = E_{L_p}\left[\lceil L_p \rceil \cdot P_m(1 - P_m)^{\lceil L_p \rceil - 1}\right] \approx l_p \cdot P_m. \tag{2.27}$$

Thus, we have

$$P_k^{C_2} \approx \Pr\{N_m = 1\} \cdot P_k^{C_2,1} \approx P_m \cdot (l_p - 1) \cdot P_k^T(0). \tag{2.28}$$

Next, we derive the $P_k^{C_1}$, which has a relationship with the probability of the false alarm P_f. The probability that events occur n times during the idle period starting at the slot k is written as

$$\Pr\{N_f = n\} = E_{L_i}\left[\binom{\lceil L_i \rceil}{n} P_f^n (1 - P_f)^{\lceil L_i \rceil - n}\right], \tag{2.29}$$

where L_i is the number of SU slots that a PU idle period spans over, and its expectation $E[L_i] = l_i, (l_i \in \mathbf{R})$, N_f is the number of false alarm events during an idle period. Let $P_k^{C_1,n}$ denote the type-I collision probability when false alarm events occur n times. Therefore, we have

$$P_k^{C_1} = \sum_{n=0}^{+\infty} \Pr\{N_f = n\} \cdot P_k^{C_1,n}. \tag{2.30}$$

Similarly, it is obvious that $\Pr\{N_f = n\}$ is negligible when P_f is small enough (e.g., $P_f \leq 0.05$) and $n \geq 2$. Therefore, we only need to consider the case of $n \leq 1$. Obviously,

$$P_k^{C_1} \approx \Pr\{N_f = 0\} \cdot P_k^{C_1,0} + \Pr\{N_f = 1\} \cdot P_k^{C_1,1}, \tag{2.31}$$

where

$$P_k^{C_1,0} = \sum_{i=0}^{N-1} B_k(i) \cdot P_k^T(i), \tag{2.32}$$

Table 2.2   Numerical Calculation of Collision Probability and Throughput under Sensing Errors

	P_f	P_m	P_k^C	T_k
Set-1	0	0	0.1009	0.1989
Set-2	0	0.01	0.1309 (+29.7%)	0.1989 (+0%)
Set-3	0.01	0	0.1120 (+11%)	0.2045 (+2.82%)

$$P_k^{C_1,1} = \sum_{i=1}^{N-1} \left[\frac{1}{i+1} \cdot B_k(i) \cdot \sum_{n=0}^{i-1} P_k^T(n) \right]. \qquad (2.33)$$

Appendix 2-B shows the derivation of $P_k^{C_1,1}$ in detail.

Second, we quantify the impact of sensing errors on the SU throughput. When missed detection happens, the SU may transmit even when the channel is occupied by the PU. However, if the SU transmission fails due to the collision, the SU throughput will remain the same. As a result, we only need to consider the impact of the false alarm on the SU throughput. The SU throughput under sensing errors is expressed as

$$T_k \approx \Pr\{N_f = 0\} \cdot T_k^0 + \Pr\{N_f = 1\} \cdot T_k^1, \qquad (2.34)$$

$$T_k^0 = \frac{\sum_{i=0}^{N-1} I_k(i) \cdot P_k^T(i)}{\sum_{i=0}^{N-1} I_k(i)}, \qquad (2.35)$$

$$T_k^1 = \frac{\sum_{i=1}^{N-1} \left\{ \frac{1}{i+1} \cdot B_k(i) \cdot \sum_{n=0}^{i-1} \left[(2i-1-2n) \cdot P_k^T(n) \right] \right\}}{\sum_{i=0}^{N-1} I_k(i)}, \qquad (2.36)$$

where T_k^n is the normalized SU throughput when false alarm events occur n times. The derivation of T_k^1 is presented in Appendix 2-C.

Table 2.2 shows different P_k^C and T_k when the set of $\{P_f, P_m\}$ is different, in which the APCR threshold $R_{TH} = 0.1$, $l_p = 4$ slots, and $l_i = 8$ slots. For Set-2, in which only missed detection occurs, the resulting APCR exceeds the predefined threshold by about 29.7% while the throughput remains unchanged. For Set-3, in which only the false alarm occurs, the resulting APCR exceeds the predefined threshold by about 11% while the throughput is greater than that of the perfect sensing case (Set-1) for 2.82%. The results show that the PSA scheme is sensitive to the sensing errors, especially missed detection. Thus, the stricter constraint is required to be imposed on the missed detection probability than the false alarm probability to constrain the APCR below a preset threshold.

2.4.2 Impact of Unknown Primary User Idle Period Distribution

We extend the PSA scheme to some more challenging situations. In a practical CR-OSAN, the SU does not have *a priori* knowledge of the distribution of the PU's idle period. Even if the type of distribution is known, the parameters of the idle period distribution might be time-varying. Therefore, we introduce a practical method to predict $I_k(i)$ and $B_k(i)$ in (2.4) and (2.5) via the hidden Markov model (HMM)-based predictor, which also enables the PSA scheme to achieve throughput improvement as well as constrain the APCR when spectrum sensing is imperfect. In Akbar and Tranter [29], HMM is adopted to predict the next PU state in the future; in this chapter, we adopt HMM to predict future PU states in N steps. The predictor learns the latest sensing history vector $\mathbf{Z}(k) = [X(k-W+1), X(k-W+2), ..., X(k)]$ that contains W ($W > 0$) sensing results from the $(k-W+1)$-th slot to the k-th slot, and then predicts the future PU states.

A discrete time HMM with M hidden states and K output symbols is a doubly embedded stochastic process [30], which is denoted by parameters set $\xi = \{\mathbf{P}, \mathbf{B}, \pi\}$, where $\mathbf{P}$ is the M-by-M state transition matrix, which denotes the underlying stochastic process that is not observable. $\mathbf{B}$ is the M-by-K output symbol probability matrix, which stands for the stochastic process that produce the output sequence observed, and the initial state probability vector π gives the probability of being in a particular state at the beginning of the process. In the context of sensing the channel occupancy by the PU, the output symbol ranges between 0 and 1, where 0 represents an idle sensing output and 1 represents a busy sensing output. It was validated in Ghosh et al. [31] that HMM is an accurate model for spectrum sensing output when the sensing is prone to errors in the form of missed detection and the false alarm. In this section, we adopt HMM to predict the future channel states based on the sensing history vector $\mathbf{Z}(k)$ and then introduce a method to use these predictions in the PSA scheme.

Generally speaking, when $\mathbf{Z}(k)$ is given, an optimal ξ to maximize the expectation of $\Pr(\mathbf{Z}(k)|\xi)$ can be trained via the Baum-Welch algorithm (BWA) [32]. Furthermore, when we get ξ, we can compute $\Pr(\mathbf{Z}(k)|\xi)$ via the forward-backward procedure [30], then we predict the probability of channel occupancy starting from slot $k+i$ by calculating $\Pr(X(k+1) = 0, ..., X(k+i) = 0, X(k+i+1) = 1, \mathbf{Z}(k)|\xi)$. Therefore, based on the trained HMM, the predicted $B_k(i)$ and $I_k(i)$ can be obtained as follows

$$\hat{B}_k(i) = \begin{cases} \frac{\Pr(X(k+i+1)=1, \mathbf{Z}(k)|\xi)}{\Pr(\mathbf{Z}(k)|\xi)}, & i = 0, \\[2ex] \frac{\Pr(X(k+1)=0, ..., X(k+i)=0, X(k+i+1)=1, \mathbf{Z}(k)|\xi)}{\Pr(\mathbf{Z}(k)|\xi)}, \\[1ex] 0 < i \leq N-1, \end{cases} \quad (2.37)$$

$$\hat{I}_k(i) = \begin{cases} \frac{\Pr(X(k+i+1)=0, \mathbf{Z}(k)|\xi)}{\Pr(\mathbf{Z}(k)|\xi)}, & i = 0, \\[2ex] \frac{\Pr(X(k+1)=0, ..., X(k+i)=0, X(k+i+1)=0, \mathbf{Z}(k)|\xi)}{\Pr(\mathbf{Z}(k)|\xi)}, \\[1ex] 0 < i \leq N-1. \end{cases} \quad (2.38)$$

Then, the solution of the optimal transmission probabilities in (2.6) is obtained. For the PSA scheme based on HMM, the HMM parameter set ξ can be updated continuously or periodically using the latest sensing results $\mathbf{Z}(k)$ as the training sequence. As a result, the SU has the capability to track the variation of the PU's idle period pattern.

2.4.3 Performance Comparisons

In the following, we present simulation results of the PSA scheme. In the simulations, the slot duration of the SU is $L = 60$ ms. For PU traffic, the PU packet length is fixed to 4 slots, while the PU idle period follows exponential distribution or hyper-Erlang distribution. For exponential distribution, rate parameter $\lambda = \frac{1}{8}$ slot^{-1}; for the hyper-Erlang distribution, the distribution parameters are given in Table 2.1. For the PSA scheme, the number of prediction steps $N = 40$, the hidden state number of the HMM-based predictor $M = 16$, and the training sequence length $W = 3000$.

Performance of the schemes listed below are simulated and compared in this chapter.

■ *Baseline*—Performance of the baseline scheme

■ *PSA (theoretical)*—Theoretical performance of the PSA scheme through numerical calculation with (2.6)

■ *PSA (simulation)*—Performance evaluated in simulation, under the assumption that the SU has *a priori* knowledge of the PU idle period distribution

■ *PSA (HMM)*—Performance evaluated in simulation, using the HMM-based predictor to predict the PU traffic pattern that is unknown to the SU

First, we compare the APCR perceived by the PU when the SU adopts different schemes. A group of APCR thresholds $R_{TH} \in [0, 0.2]$ is given and the resulting APCR is measured. We select the low region out of all possible APCR thresholds since over-loose APCR thresholds (e.g., $R_{TH} > 0.2$) may not constrain the SU for sufficient protection over the PU. Then, the measured APCR versus APCR threshold curves for exponential distribution and hyper-Erlang distribution are plotted in Figures 2.5 and 2.6, respectively. Apparently, the baseline scheme constrains the APCR perfectly. The PSA scheme using the HMM-based predictor or *a priori* knowledge can achieve a performance very close to that of the baseline scheme.

Next, the normalized throughput of the SU is compared for the two schemes. The normalized throughput versus the APCR threshold curves for exponential distribution and hyper-Erlang distribution are plotted in Figures 2.7 and 2.8, respectively. The normalized throughput achieved by the baseline scheme equals to the APCR threshold. It is observed that the PSA scheme outperforms the baseline scheme significantly when the distribution of the PU idle period is a hyper-Erlang distribution. When the APCR threshold is set to 0.1, the normalized throughput of the PSA scheme is about 0.2, while the normalized throughput of the baseline scheme is

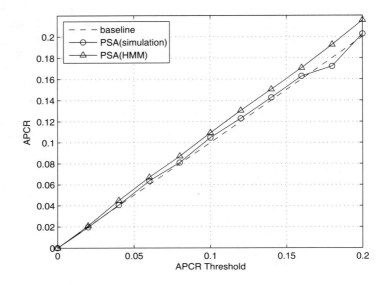

Figure 2.5: The APCR perceived by the PU with exponential distribution.

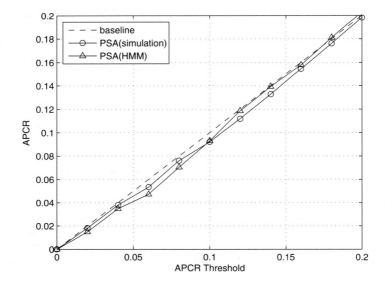

Figure 2.6: The APCR perceived by the PU with hyper-Erlang distribution.

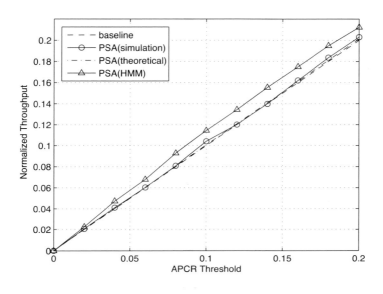

Figure 2.7: Normalized throughput of the SU with exponential distribution.

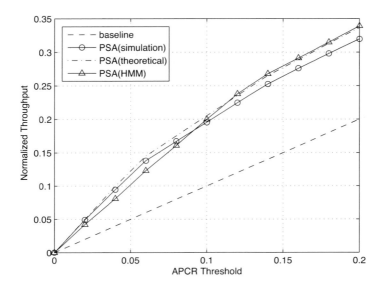

Figure 2.8: Normalized throughput of the SU with hyper-Erlang distribution.

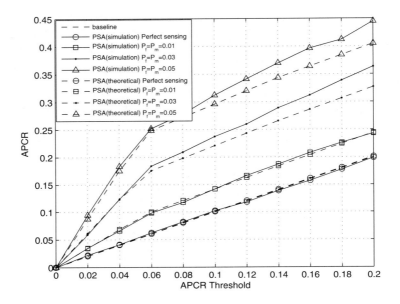

Figure 2.9: The APCR perceived by the PU with imperfect sensing.

only 0.1. By contrast, the PSA scheme cannot achieve higher throughput than the baseline scheme for exponential distribution, which is consistent with (2.19). The PSA scheme that predicts unknown future channel states via the HMM-based predictor achieves a performance very close to the scheme that uses *a priori* knowledge of the PU idle period distribution. Meanwhile, the expectation of the throughput calculated with (2.5) corresponds with the measured throughput in the Monte Carlo simulation. These results verify that the PSA scheme can increase the SU throughput significantly when the distribution of the PU idle period is a hyper-Erlang distribution.

In the following, the impact of sensing errors on the APCR and the SU throughput for the PSA scheme are presented. For comparison, three sets of sensing error probabilities $\{P_f, P_m\}$ are simulated ($\{0.01, 0.01\}$, $\{0.03, 0.03\}$, and $\{0.05, 0.05\}$). When the SU has *a priori* knowledge of the PU idle period distribution, the APCR versus APCR threshold curves and normalized throughput versus APCR threshold curves are plotted in Figures 2.9 and 2.10, respectively. Results show that, for the APCR and SU throughput under sensing errors, the derived results are consistent with the simulated ones. With the increase of P_f and P_m, the gaps between the derived curves and the simulated curves become bigger. The reason lies in the fact that P_k^C and T_k are approximately derived under the assumption that P_f and P_m are minor. It is observed that the sensing errors will cause serious consequences, that is, the APCR exceeds the threshold significantly ($> 40\%$) even when the sensing error

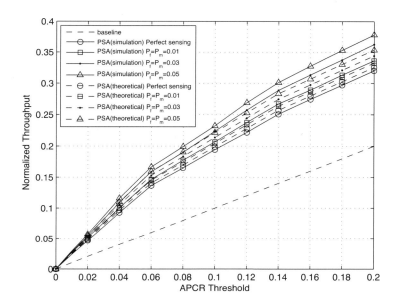

Figure 2.10: Normalized throughput of the SU with imperfect sensing.

probabilities are generic ($P_f = P_m = 0.01$). The SU throughput is boosted mainly resulting from the false alarm. It verifies that the PSA scheme that uses *a priori* knowledge of the PU idle period distribution is sensitive to the sensing errors. When the SU adopts the HMM-based predictor to predict the PU traffic pattern that is unknown to it, the APCR versus APCR threshold curves and the normalized throughput versus APCR threshold curves are plotted in Figure 2.11 and Figure 2.12, respectively. Results show that the APCR is not increased much even when the sensing errors become severe, while the SU can gain considerable throughput improvement with the PSA scheme with the HMM-based predictor than the baseline scheme in this situation. The reason lies in the capability of the HMM-based predictor, which can learn and adapt to the traffic pattern with imperfect sensing results.

Finally, we test the PSA scheme with dynamic PU traffic parameters, in which the PU idle period is a hyper-Erlang distribution with time-varying rate parameters, that is, $\lambda_1 = \left[0.2 + 0.1 \cdot \cos\left(\frac{2\pi}{24} \cdot t\right) \right]$ slot^{-1}, $\lambda_2 = \left[0.1 + 0.05 \cdot \cos\left(\frac{2\pi}{24} \cdot t\right) \right]$ slot^{-1}, where time t is in hours. We make a performance comparison between our PSA scheme and the adaptive opportunistic spectrum access (AOSA) scheme in Huang, Liu, and Ding [24]. The AOSA scheme aims to maximize the SU throughput under the collision probability constraint without *a priori* knowledge of the PU traffic pattern. The main idea of the AOSA scheme is to successively transmit Q SU data sub-slots during a PU idle period. By calculating the resulting APCR $\tilde{R}(j)$ in the interval j, $j = 1, 2, \cdots$, the Q in the interval $j + 1$ is updated as follows

$$Q(j+1) = Q(j) + \beta(j)Q(0)\frac{R_{\text{TH}} - \tilde{R}(j)}{R_{\text{TH}}}. \qquad (2.39)$$

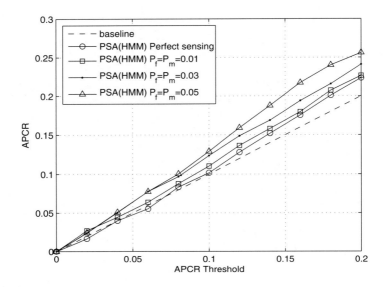

Figure 2.11: The APCR perceived by the PU with imperfect sensing.

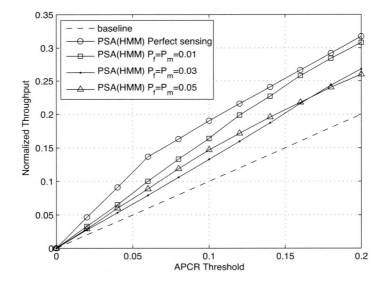

Figure 2.12: Normalized throughput of the SU with imperfect sensing.

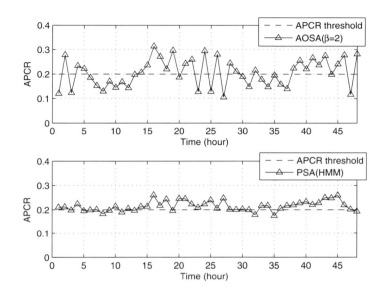

Figure 2.13: The APCR perceived by the PU during a duration of 48 hours.

Where $Q(0)$ is the initial valve of Q, which is set to the optimal Q for exponential PU idle period distribution, that is, $Q(0) = -l_i \ln(1 - R_{TH})$; $\beta(j)$ denotes the step size; R_{TH} denotes the preset APCR threshold. In the simulation, the interval is set to 60 minutes (18000 slots), that is, parameter Q is updated every 60 minutes. The step size $\beta(j) = \beta = 2$, $j = 1, 2, \cdots$.

It is assumed that the PU traffic pattern is unknown to both schemes. For the PSA scheme using the HMM-based predictor, the HMM parameter set ξ is updated every 60 minutes and the training sequence length is 3000. The APCR threshold $R_{TH} = 0.2$. The APCR and the SU throughput are measured every 60 minutes. The APCR versus time curves are plotted in Figure 2.13; it is observed that the APCR with the PSA scheme is well constrained by the APCR threshold, while the APCR with the AOSA scheme jitters around the APCR threshold dramatically. The SU throughput versus time curves are plotted in Figure 2.14. We find that the PSA scheme achieves a SU throughput very close to the theoretical optimal value. The results in the dynamic experiment verify that the HMM-based predictor achieves better performance in the tracking of idle period distribution variation compared with the AOSA scheme in Huang, Liu, and Ding [24].

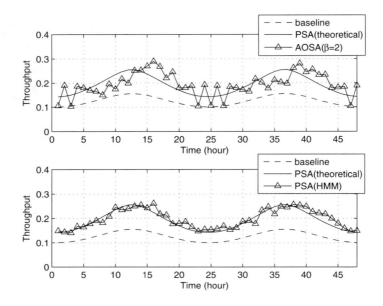

Figure 2.14: The throughput of the SU during a duration of 48 hours.

2.5 Summary

In this chapter, a PSA scheme has been introduced to maximize the SU throughput while constraining the APCR perceived by the PU under a required threshold. Theoretical analysis and numerical results show that, when the PU idle period follows the hyper-Erlang distribution, significant throughput improvement is achieved by adopting the PSA scheme. When the spectrum sensing at the SU is imperfect, we have analyzed and quantified the impact of the sensing errors on the SU performance with the PSA scheme. Numerical results show that the performance of the PSA scheme is sensitive to the sensing errors, especially missed detection. For more practical situations, in which the idle period distribution of the PU and its parameters are unknown to the SU, as well as the spectrum sensing at the SU is imperfect, the HMM-based predictor enables the PSA scheme to achieve considerable throughput improvement and satisfactory APCR constraint.

Appendix 2-A: Derivation of T_k

From the definitions of $B_k(i)$ and $I_k(i)$, we have

$$I_k(i) = \sum_{n=i+1}^{+\infty} B_k(n). \tag{2.40}$$

When N is large enough (but finite), the expected number of successfully utilized idle slots is

$$
n_k^S = \sum_{i=1}^{N} \left[B_k(i) \cdot \sum_{n=0}^{i-1} P_k^T(n) \right] = \sum_{i=0}^{N-1} \left[P_k^T(i) \cdot \sum_{n=i+1}^{N} B_k(n) \right]
$$

$$
= \sum_{i=0}^{N-1} I_k(i) \cdot P_k^T(i),
$$

(2.41)

the expectation of the idle period length is

$$
l_i = \sum_{i=1}^{N} i \cdot B_k(i)
$$

$$
= \sum_{n=1}^{N} B_k(n) + \sum_{n=2}^{N} B_k(n) + \ldots + \sum_{n=N}^{N} B_k(n)
$$

$$
= \sum_{i=0}^{N-1} I_k(i),
$$

(2.42)

where $l_i \in \mathbf{R}$ is the expected number of SU slots that a PU idle period covers.

Thus, we have the normalized throughput as

$$
T^k = \frac{n_k^S}{l_i} = \frac{\displaystyle\sum_{i=0}^{N-1} I_k(i) \cdot P_k^T(i)}{\displaystyle\sum_{i=0}^{N-1} I_k(i)}.
$$

(2.43)

Appendix 2-B: Derivation of $P_k^{C_1,1}$

The type-I collision occurs at the data sub-slot that the primary user returns to the channel. Suppose that the idle period covers i idle slots (contains $i+1$ sensing sub-slots) with probability $B_k(i)$, $(i > 0)$, the probability of a false alarm in a certain sensing sub-slot is $\frac{1}{i+1}$, thus the expectation of collision numbers with a certain idle period length L_i is

$$
E[N_k^C | L_i] = \frac{1}{i+1} \cdot \sum_{n=0}^{i-1} P_k^T(n),
$$

(2.44)

then, we have

$$
P_k^{C_1,1} = E_{L_i} \left[E[N_k^C | L_i] \right]
$$

$$
= \sum_{i=1}^{N-1} \left[\frac{1}{i+1} \cdot B_k(i) \cdot \sum_{n=0}^{i-1} P_k^T(n) \right].
$$

(2.45)

Appendix 2-C: Derivation of T_k^1

The expected number of successfully utilized idle slots with a certain idle period length L_i is

$$
\begin{aligned}
E[N_k^S | L_i] &= \frac{1}{i+1} \cdot \left\{ \sum_{n=0}^{i-1} P_k^T(n) + 1_{[i=2]} \cdot 2 \cdot P_k^T(0) + 1_{[i>2]} \cdot \right. \\
&\quad \left. \left[2 \cdot \sum_{n=0}^{i-2} P_k^T(n) + \sum_{m=1}^{i-2} \left(\sum_{u=0}^{m-1} P_k^T(u) + \sum_{v=0}^{i-2-m} P_k^T(v) \right) \right] \right\} \\
&= \frac{1}{i+1} \cdot \sum_{n=0}^{i-1} \left[(2i - 1 - 2n) \cdot P_k^T(n) \right], \quad\quad (2.46)
\end{aligned}
$$

where $1_{[\cdot]}$ denotes the indicator function. Thus, we have

$$
\begin{aligned}
T_k^1 &= \frac{E_{L_i}\left[E[N_k^S | L_i] \right]}{l_i} \\
&= \frac{\sum_{i=1}^{N-1} \left\{ \frac{1}{i+1} \cdot B_k(i) \cdot \sum_{n=0}^{i-1} \left[(2i - 1 - 2n) \cdot P_k^T(n) \right] \right\}}{\sum_{i=0}^{N-1} I_k(i)}. \quad\quad (2.47)
\end{aligned}
$$

Chapter 3

Sensing Node Allocation in a Cognitive Radio Network with Cooperative Sensing

Spectrum sensing for acquiring the availability of the licensed spectrum band is commonly recognized as one of the most fundamental elements in cognitive radio-based networks. Spectrum sensing can be realized as a two-layer mechanism [105]. On the one hand, PHY-layer sensing focuses on efficiently detecting the signals of primary users to identify whether the primary users are present or not. Some PHY-layer sensing methods including energy detection [43], matched filter [44], and feature detection have been studied [45]. Moreover, some sequential spectrum sensing schemes have also been investigated, such as the sequential shifted chi-square test [46], which effectively reduces the average sample number compared with fixed-sample-size energy detection. On the other hand, MAC-layer sensing, which plays an important role in a cognitive radio-based centralized network, determines the channels that secondary users should sense and access in each slot for good performance in terms of sensing delays, throughput of secondary users, and so forth. Sensing delays result in performance degradation of secondary users, especially in broadband communications systems since more sensing time is needed to acquire enough spectrum opportunities. Therefore, an important issue in MAC-layer sensing is how to acquire more spectrum opportunities quickly. Collaborative spectrum sensing has also been well studied in cognitive radio-based centralized networks in recent years since the degradation of sensing performance caused by multi-path fading and shadowing can be effectively overcome via collaborative sensing. Moreover, the sensing time can also be reduced via collaborative sensing since more sensing data can be obtained simultaneously.

In multi-user multi-channel cognitive radio-based centralized networks, there are new challenges to design algorithms for the fast discovery of available channels. When the number of sensing nodes is large, some nodes may be wasted if all the sensing nodes collaboratively sense one channel, since the observations from only a part of the sensing nodes are needed to determine the channel state with the predefined sensing performance guarantee. Therefore, we hope to optimally allocate the resources of sensing nodes for fast discovery of available channels. In this chapter, we aim to acquire more available channels in a short sensing time based on the sequential probability ratio test (SPRT) theory. Specifically, we dynamically select the number of secondary users based on the SPRT theory to sense channels in each slot for fast acquisition of spectrum opportunities. A MAC-layer sensing framework is constructed for efficient acquisition of spectrum opportunities in a cognitive radio network with cooperative sensing [52]. In the framework, we employ the sequential probability ratio test and develop a new adaptive collaboration sensing scheme for multi-users to collaborate during multiple slots, in which we effectively utilize the resources of secondary users to sense the channels for fast acquisition of spectrum opportunities. Furthermore, the scheme is formulated as an optimization problem and we derive the optimal solution with low complexity, which is based on dynamic programming theory.

3.1 Multi-User Multi-Channel System Model

The scenario considered in this chapter is illustrated in Figure 3.1. We use the concept of a cognitive cell, which consists of a cognitive base station (CBS) and a group of M SUs. In the range of a cognitive cell, the CBS can effectively receive the data from the SUs and schedule SUs for spectrum sensing. We only consider collaborative sensing among all the SUs in this chapter. In practice, the role of these collaborative SUs can also be replaced by sensor networks [53]. The licensed spectrum band is equally divided into N non-overlapping narrowband data channels, and all channels are under Rayleigh fading. The goal of spectrum sensing is to decide between the following two hypotheses

$$x_{ij}(t) = \begin{cases} v_{ij}(t), & H_{i0} \\ h_{ij} \cdot s_i(t) + v_{ij}(t), & H_{i1} \end{cases} \tag{3.1}$$

where $x_{ij}(t)$ is the signal received by the j-th SU in the i-th channel, and $i = 1, 2, \cdots, N$, $j = 1, 2, \cdots, M$. $s_i(t)$ is the primary user's transmitted signal in the i-th channel, which is assumed with a mean zero and variance σ_{is}^2, $v_{ij}(t)$ is the complex additive white Gaussian noise (AWGN) with a mean zero and variance σ_{ijv}^2, h_{ij} is the instantaneous channel gain between the primary user in the i-th channel and the j-th SU. For mathematical brevity, we assume that $\sigma_{is}^2 = \sigma_s^2$ and $\sigma_{ijv}^2 = \sigma_v^2$ for $i = 1, 2, \cdots, N$, $j = 1, 2, \cdots, M$, and $h_{ij} = h_j$ for $i = 1, 2, \cdots, N$. Obviously, all these

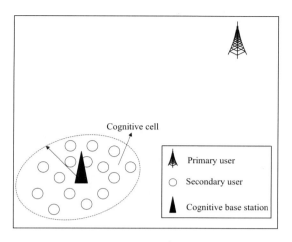

Figure 3.1: The considered scenario.

assumptions are reasonable when we consider that the licensed spectrum bands are located in the downlink bands of a cellular system or the TV and broadcasting bands. In addition, H_{i0} is the null hypothesis, which represents that the primary user is inactive in the i-th channel. H_{i1} is the alternative hypothesis, which represents that the primary user is active in the i-th channel.

We employ an energy detector in each SU to determine whether the primary user is active or not, which is a simple and effective method for the detection of unknown signals. In each energy detector, the received signal is fed to a band-pass filter with the center frequency f_c and the bandwidth of interest W. This filter is followed by a squaring device to measure the received energy and an integrator to determine the observation interval T. With the Nyquist theorem, a signal of duration T and bandwidth W should have $S \geqslant 2TW$ samples. Then, $t_{ij} = \frac{1}{S}\sum_{s=1}^{S}|x_{ij}(s)|^2$ is defined as one observation from the j-SU on channel i. Although t_{ij} has a chi-square distribution [44, 54], according to the central limit theorem, t_{ij} is asymptotically normally distributed if S is large enough ($S \geqslant 20$ is often sufficient in practice) [55]. Specifically, for large S, we can model the statistic of t_{ij} as follows for analytical simplicity

$$t_{ij} \sim \begin{cases} \mathrm{N}\left(\sigma_v^2, \frac{\sigma_v^4}{S}\right), & H_{i0} \\ \mathrm{N}\left(\sigma_v^2(1+\gamma_j), \frac{\sigma_v^4(1+2\gamma_j)}{S}\right), & H_{i1} \end{cases} \tag{3.2}$$

where $\gamma_j = \frac{h_j^2 \sigma_s^2}{\sigma_v^2}$ is the received signal-to-noise ratio (SNR) of the j-th SU.

In practice, it is shown in Figure 3.2 that the observation interval T must be large enough for a low probability of a false alarm, denoted as P_{fa}, and a high probability of detection, denoted as P_d [20]. The large observation interval T cannot satisfy the requirement of fast spectrum opportunity acquisition in cognitive radio networks

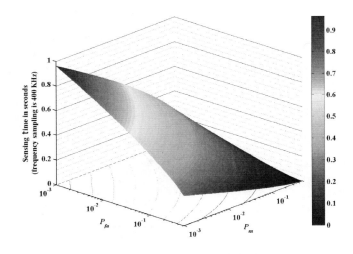

Figure 3.2: Trade-off between sensing time and detection performance (P_m versus P_{fa}).

(CRNs). Hence, in this chapter, we introduce a novel adaptive collaboration sensing scheme for an efficient acquisition of spectrum opportunities in a limited sensing time.

We employ a slot-based sensing method. One SU provides one observation on a certain channel in each slot, that is, the number of observations is equal to the number of SUs that sense channels in each slot. The system model is illustrated in Figure 3.3. A CBS consists of three modules: the fusion center, the sensing schedule module, and the available channel list. The fusion center obtains all the observations from the SUs and determines the state of the channels. Then, the idle channels are added to the available channel list, the busy channels are dropped, and the sensing results

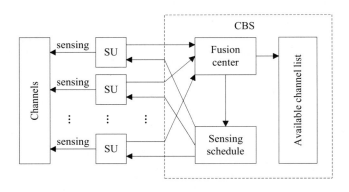

Figure 3.3: The system model.

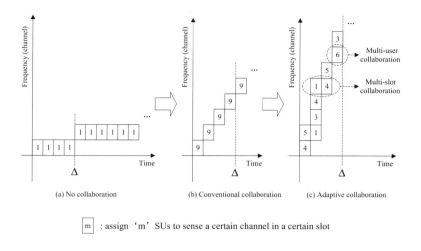

(a) No collaboration   (b) Conventional collaboration   (c) Adaptive collaboration

| m | : assign 'm' SUs to sense a certain channel in a certain slot

Figure 3.4: The basic idea of an adaptive collaboration sensing scheme (ACSS).

of the other channels that need more observation are fed into the sensing schedule module. After the schedule algorithm, the sensing schedule module employs the SUs to sense the appropriate channels in the next slot. This process is repeated until all the channels are determined. Moreover, since primary users may appear at any time, CBS can be initialized and operates the sensing process discussed above once again. This chapter focuses on designing the sensing schedule algorithm in our MAC-layer sensing framework.

3.2 Adaptive Collaboration Sensing Scheme

In this section, the details of our scheme are given. First, the basic idea of our scheme is briefly illustrated. Then, after introducing the details of the collaboration based on the SPRT method, we introduce the adaptive collaboration sensing scheme and formulate the problem as an optimization problem.

3.2.1 Basic Idea

To reduce the sensing time while maintaining the performance of P_{fa} and P_m, that is, maintaining the number of observations, we consider a novel adaptive collaboration sensing scheme (ACSS) based on the multi-user multi-slot SPRT method in the CBS. Figure 3.4 is a schematic diagram that illustrates the basic idea of ACSS. In Figure 3.4, the number in the box means the number of SUs that we assign to sense a certain channel in a certain slot. When no collaboration is adopted, only one observation on a certain channel is obtained in one slot, which is shown in Figure 3.4a, and until the Δ-th slot, only one channel's state is determined. The conventional collaboration is shown in Figure 3.4b, all the SUs (in this schematic diagram, we assume the number of SUs $M = 9$) sense one channel in one slot, that is, M observations on a certain channel are obtained in one slot, and until the Δ-th slot, four channel states are determined. When the number of SUs is large, some observations from the SUs may be wasted in the conventional collaboration scheme since only a portion of the observations are needed to determine the channel state with a certain P_{fa} and P_d guarantee. To solve this problem, we introduce the adaptive collaboration sensing scheme (ACSS) which is shown in Figure 3.4c. The appropriate number of observations on a certain channel is obtained in one slot, and thus more than one channel's state will be determined in one slot. From Figure 3.4c, eight channel states are determined until the Δ-th slot. Obviously, the key issue to be solved is how many SUs should collaborate to detect the signals on channel i in each slot, $i = 1, \cdots, N$. Before introducing our algorithm, some details of the collaboration based on the SPRT method will be explained in the following section.

3.2.2 Sequential Probability Ratio Test

In the current slot, we assume each SU observes a certain channel and obtains one observation on this channel. All observations are assumed to be independent and identically distributed (i.i.d.). Let $\mathbf{t}_i = [t_{i1} \ t_{i2} \ \cdots \ t_{iU_i}]$ denote the observations on the channel i, where U_i is the number of observations (i.e., the number of SUs that sense the channel i), $0 \leqslant U_i \leqslant M$, $i = 1, \cdots, N$, and $\sum_{i=1}^{N} U_i = M$. The likelihood ratio of $\mathbf{t}_i$ is given by [47]

$$\Lambda(\mathbf{t}_i) = \prod_{k=1}^{U_i} \Lambda(t_{ik}) = \prod_{k=1}^{U_i} \frac{f(t_{ik}|H_{i1})}{f(t_{ik}|H_{i0})}, \qquad (3.3)$$

where $f(t_{ik}|H_{i0})$ is the probability density function of the normal distribution $N\left(\sigma_v^2, \frac{\sigma_v^4}{S}\right)$ in (3.2), and $f(t_{ik}|H_{i1})$ is the probability density function of the normal distribution $N\left(\sigma_v^2(1+\gamma_k), \frac{\sigma_v^4(1+2\gamma_k)}{S}\right)$ in (3.2).

The log-likelihood ratio (LLR) of $\Lambda(\mathbf{t}_i)$ is obtained by

$$L(\mathbf{t}_i) = \ln \Lambda(\mathbf{t}_i) = \sum_{k=1}^{U_i} \ln \Lambda(t_{ik}) = \sum_{k=1}^{U_i} L(t_{ik}), \qquad (3.4)$$

where $L(t_{ik})$ is the LLR of the k-th observation on the i-th channel in the current slot, and the expression $L(t_{ik})$ is given in Appendix 3-A. If T_{ik} denotes the statistic of $L(t_{ik})$, we have

$$T_{ik} \sim \begin{cases} \chi_1^2\left(s_{0i}^2\right) + \text{const}, & H_{i0} \\ \chi_1^2\left(s_{1i}^2\right) + \text{const}, & H_{i1} \end{cases} \qquad (3.5)$$

where s_{0i}^2 and s_{1i}^2 are the non-central parameters under H_{i0} and H_{i1}, respectively.

Then, the LLRs of observations from all the SUs are transmitted to the fusion center, and decisions for utilizing the spectrum opportunities are made via the SPRT method. The decision rule for SPRT with thresholds η_0 and η_1, denoted as

$SPRT(\eta_0, \eta_1)$, is given by

$$\begin{cases} L(\mathbf{t}_i) \geqslant \ln \eta_1, & \text{accept } H_{i1} \\ L(\mathbf{t}_i) \leqslant \ln \eta_0, & \text{accept } H_{i0} \\ \ln \eta_1 \geqslant L(\mathbf{t}_i) \geqslant \ln \eta_0, & \text{take another observation} \end{cases} \tag{3.6}$$

The following properties of SPRT are well-known in Wald [56].

Theorem 3.1
Let $P_m = \alpha$ and $P_{fa} = \beta$ be the probabilities associated with $SPRT(\eta_0, \eta_1)$. Then, the two thresholds η_0 and η_1 satisfy

$$\eta_1 \leqslant \tfrac{1-\alpha}{\beta}, \quad \eta_0 \geqslant \tfrac{\alpha}{1-\beta}. \tag{3.7}$$

When a decision to accept H_{i1} or H_{i0} is made, if the LLR is exactly equal to the corresponding threshold, which happens if the LLR is a continuous process, the above inequalities become equalities. In practice, we usually employ the equalities, which are excellent approximations in many cases [47]. Moreover, we have the following theorem [56] to guarantee the practical probability of missed detection and probability of a false alarm.

Theorem 3.2
Let $\eta_1 = \tfrac{1-\alpha}{\beta}$ and $\eta_0 = \tfrac{\alpha}{1-\beta}$. Let α^* denote the practical probability of missed detection and β^* denote the practical probability of a false alarm. Then, if $\alpha + \beta < 1$, we have:

$$\alpha^* \leqslant \tfrac{\alpha}{1-\beta}, \quad \beta^* \leqslant \tfrac{\beta}{1-\alpha}. \tag{3.8}$$

3.2.3 Optimal Sensing Node Allocation

To efficiently acquire available channels in a limited sensing time, we hope to assign an appropriate number of SUs to sense channels in each slot so that the most available channels are acquired until the Δ-th slot, $\Delta = 1, 2, \cdots, +\infty$. Let $\mathbf{U}_\delta$ denote the assignment vector in the δ-th slot, and the element of $\mathbf{U}_\delta$, $U_{i\delta}$ denotes the number of SUs assigned to channel i in slot δ, where $i = 1, 2, \cdots, N$, $1 \leqslant \delta \leqslant \Delta$. In addition, let $N_{\Delta,available}$ denote the expectation of the number of available channels acquired until the slot Δ under certain assignment vectors $\mathbf{U}_1, \cdots, \mathbf{U}_\Delta$. We hope to find the optimal assignment set $\mathbf{U}^* = \{\mathbf{U}_1, \cdots, \mathbf{U}_\Delta\}$ such that the corresponding $N^*_{\Delta,available}$ is optimal, that is, $\forall \mathbf{U}^\dagger \neq \mathbf{U}^*$, $N^\dagger_{\Delta,available} < N^*_{\Delta,available}$.

However, in the assignment matrix $\mathbf{U}^*$, the δ-th column vector, that is, the assignment vector in slot δ, depends heavily on past observation results, which are based on assignment vectors in slots $1, 2, \cdots, (\delta - 1)$, that is, the $1, 2, \cdots, (\delta - 1)$-th columns of $\mathbf{U}^*$. Obviously, there are causal relationships between the columns in assignment matrix and the optimal assignment matrix $\mathbf{U}^*$ cannot be obtained directly. Hence, it is necessary to obtain $\mathbf{U}^*$ slot-by-slot according to the past observation results.

Obviously, we have $N^*_{1,available} = n^*_{1,available}$ when $\Delta = 1$ and $N^*_{\delta,available} = N^*_{\delta-1,available} + n^*_{\delta,available}$ when $\Delta > 1$, where $n^*_{\delta,available}$ denotes the expectation of the optimal number of available channels acquired in slot δ. Since in slot δ, $N^*_{\delta-1,available}$ has been obtained and fixed, we maximize $n^*_{\delta,available}$ instead of $N^*_{\delta,available}$ in the δ slot. For mathematical brevity, the index δ is omitted in the following. $n_{available}$ is obtained by $\sum_{i=1}^{N} G_i(U_i, i)$, where $G_i(U_i, i)$ is the probability of channel i being available when U_i SUs sense channel i. Based on (5.4), $G_i(U_i, i)$ can be obtained by

$$G_i(U_i, i) = \begin{cases} P\left(\sum_{k=1}^{U_i} T_{ik} < \ln \eta_0 - \xi_i\right), & U_i > 0 \\ 0, & U_i = 0 \end{cases} \tag{3.9}$$

where ξ_i denotes the LLR of the observations until the current slot on the channel i. $P\left(\sum_{k=1}^{U_i} T_{ik} < \ln \eta_0 - \xi_i\right)$ is derived in Appendix 3-B in detail.

Based on the discussions above, we formulate this problem as the following optimization problem

$$\max_{\mathbf{U}} \left\{ \sum_{i=1}^{N} G_i(U_i, i) \right\},$$

$$s.t. \begin{cases} \sum_{i=1}^{N} U_i \leqslant M \\ 0 \leqslant U_i \leqslant M \end{cases}, \tag{3.10}$$

where $\mathbf{U} = [U_1, \cdots, U_i, \cdots, U_N]$. Note that the thresholds for the SPRT chosen according to Theorem 3.2 are fixed, and the only parameters that are being optimized are the components of matrix $\mathbf{U}$.

It is obvious that this is a finite state variable optimization problem. The optimal solution will be obtained when N^M possible $\mathbf{U}$ that satisfy (3.10) are searched. However, if assuming that calculating $\sum_{i=1}^{N} G_i(U_i, i)$ for one concrete $\mathbf{U}$ has the complexity of $O(1)$, the complexity of this brute-force search is $O(N^M)$. Fortunately, after some transformation, the optimal solution of the optimization problem is obtained based on dynamic programming theory with low complexity.

In general, dynamic programming is an effective method to solve complex problems. Based on the principle of optimality (a.k.a. the principle of contradiction) [57], dynamic programming is applicable to problems exhibiting the properties of overlapping subproblems, which are only slightly smaller and have an optimal substructure. Actually, dynamic programming with the Viterbi algorithm as a well-known incarnation avoids an exhaustive search and makes it possible to find an optimal solution with low complexity.

We now explain how to solve the optimization problem (3.10) based on dynamic programming theory. First, to establish the dynamic programming formulation, we define a variable x_i, which denotes the number of the left SUs that senses channels $i, i+1, \cdots, N$. Obviously, we have the relationship

$$\begin{cases} x_{i+1} = x_i - U_i \\ x_1 = M \end{cases}. \tag{3.11}$$

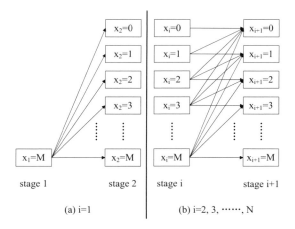

Figure 3.5: The complexity of the optimal solution based on dynamic programming, $U_i = x_i - x_{i+1}$. The arrow lines denote feasible solutions.

In addition, $I(x_i, i)$ denotes the optimal solution when x_i SUs senses channels $i, i+1, \cdots, N$, and $\mathbf{U}_{i,x_i} = [U_{i,x_i}, U_{i+1,x_i}, \cdots, U_{N,x_i}]$ denotes the corresponding optimal assignment vector. We then obtain the following recursive equation [58]

$$
\begin{aligned}
I(x_i, i) &= \max_{U_i, U_{i+1}, \cdots, U_N} \sum_{l=i}^{N} G_l(U_l, l) \\
&= \max_{U_i, U_{i+1}, \cdots, U_N} \left\{ G_i(U_i, i) + \sum_{l=i+1}^{N} G_l(U_l, l) \right\} \\
&= \max_{U_i} \left\{ \max_{U_{i+1}, \cdots, U_N} \left\{ G_i(U_i, i) + \sum_{l=i+1}^{N} G_l(U_l, l) \right\} \right\} \\
&= \max_{U_i} \left\{ G_i(U_i, i) + \max_{U_{i+1}, \cdots, U_N} \left\{ \sum_{l=i+1}^{N} G_l(U_l, l) \right\} \right\} \\
&= \max_{U_i} \left\{ G_i(U_i, i) + I(x_{i+1}, i+1) \right\} \\
&= \max_{U_i} \left\{ G_i(U_i, i) + I(x_i - U_i, i+1) \right\},
\end{aligned}
\tag{3.12}
$$

$$
\mathbf{U}_{i,x_i} = \arg \max_{U_i, U_{i+1}, \cdots, U_N} \sum_{l=i}^{N} G_l(U_l, l),
\tag{3.13}
$$

and the boundary conditions are

$$
\left\{
\begin{array}{l}
U_N \leqslant x_N \\
I(x_N, N) = \max_{U_N = \{0, 1, \cdots, x_N\}} \left\{ G_N(U_N, N) \right\}
\end{array}
\right. \cdot
\tag{3.14}
$$

Then, solving the optimization problem (3.10) is transformed into finding $I(x_1,1)$ and $\mathbf{U}_{1,x_1}$, which are easily obtained by applying the above recursive equation (3.12). Obviously, as illustrated in Figure 3.5, to obtain $I(x_i,i)$ and $\mathbf{U}_{i,x_i}$, $i = 2,3,\cdots,N$, we should search $(M+1)(M+2)/2$ feasible x_i and U_i, since x_i is searched from zero to M while U_i is searched from zero to x_i due to the relationship $x_i \geqslant U_i$. To obtain $I(x_1,1)$ and $\mathbf{U}_{1,x_1}$, we should search $(M+1)$ feasible x_1 and U_1, since we only search U_i from zero to M due to $x_1 = M$. Hence, the optimal solution based on dynamic programming has the complexity of $O(M^2N)$. The pseudo-code for obtaining the optimal solution based on dynamic programming is given in **Algorithm 3.1**.

Algorithm 3.1 The pseudo-code for obtaining the optimal solution based on dynamic programming.

Procedure:

1: for $i = N : -1 : 1$
2:   if $i = N$
3:     for $x_N = 0 : M$
4:       $I(x_N,N) = \max\limits_{U_N=0,\cdots,x_N} \{G_N(U_N,N)\};$
5:       $U_{N,x_N} = \arg\max\limits_{U_N=0,\cdots,x_N} \{G_N(U_N,N)\};$
6:       $\mathbf{U}_{N,x_N} = [U_{N,x_N}];$
7:     end for
8:   else if $i < N$ and $i > 1$
9:     for $x_i = 0 : M$
10:       $I(x_i,i) = \max\limits_{U_i=0,\cdots,x_i} \{G_i(U_i,i)$
11:              $+I(x_i-U_i,i+1)\};$
12:       $U_{i,x_i} = \arg\max\limits_{U_i=0,\cdots,x_i} \{G_i(U_i,i)$
13:              $+I(x_i-U_i,i+1)\};$
14:       $\mathbf{U}_{i,x_i} = [U_{i,x_i}, \mathbf{U}_{(i+1),(x_i-U_{i,x_i})}];$
15:     end for
16:   else if $i = 1$
17:     $I(x_1,1) = \max\limits_{U_1=0,\cdots,M} \{G_1(U_1,1)+I(x_2,2)\};$
18:     $U_{1,x_1} = \arg\max\limits_{U_1=0,\cdots,M} \{G_1(U_1,1)+I(x_2,2)\};$
19:     $\mathbf{U}_{1,x_1} = [U_{1,x_1}, \mathbf{U}_{2,(x_1-U_{1,x_1})}];$
20:   end if
21: end for

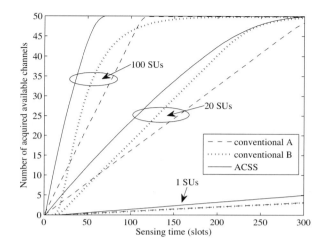

Figure 3.6: The performance of ACSS and conventional schemes with different numbers of SUs under the average SNR $\gamma = -10$ dB, $P_m = P_{fa} = 10^{-2}$.

3.3 Performance Evaluation and Analysis

In this section, we evaluate the ACSS with Monte Carlo searching. For all simulations, we consider that the channel number $N = 100$ and each channel is idle with probability $P_a = 0.5$ (i.e., about 50 channels are available), and all the channels are under Rayleigh fading. The geometric topology used in these simulations is similar to Figure 3.1. The SUs are randomly distributed within the coverage of the CBS, where the coverage radius is set to be 500 m. The path loss between the PU and SU is modeled as the Okumura-Hata propagation model [87]. We also compare it with two conventional nonadaptive collaborative schemes. In conventional scheme A, all M SUs are assigned to one channel (i.e., $\mathbf{U} = [0 \cdots M \cdots 0]$) in each slot for opportunity acquisition until the decision is obtained. In conventional scheme B, we assign one SU to sense one channel (i.e., $\mathbf{U} = [1 \cdots 1 \cdots 1]$ for $N \leqslant M$ or $\mathbf{U} = [1 \cdots 1 \cdots 1 \cdots 0]$ for $N > M$) in each slot for opportunity acquisition. In the following, we will evaluate the performance of the ACSS compared with the conventional schemes. Meanwhile, we investigate the impact of P_m, P_{fa}, SU numbers, and SNRs (different SNRs reflect the impact of varying path losses) on the performance of ACSS.

Figure 3.6 shows the performance of the ACSS compared with the conventional schemes with the number of SUs $M = 100, 20, 1$ under the average SNR $\gamma = -10$ dB, and $P_m = P_{fa} = 10^{-2}$. The horizontal axis represents the sensing time, and the vertical axis represents the number of the acquired available channels. It is obvious that an ACSS with a different number of SUs acquires the available channels faster than the conventional schemes. In addition, it is necessary to explain the simulation

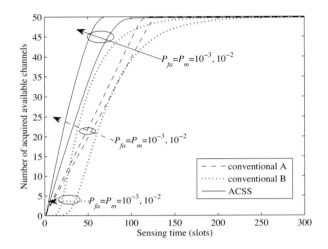

Figure 3.7: The performance of ACSS and conventional schemes with 100 SUs and different P_m, P_{fa} under the average SNR $\gamma = -10$ dB.

result when $M = 1$. In this situation, our scheme also outperforms the conventional schemes since we always sense the best channel that is available with the highest probability in the ACSS and sense a fixed channel without considering the probability of being available until the channel has been determined as idle or busy in conventional schemes. Moreover, when $M = 1$, the two conventional schemes are the same. Thus, the performance curves of these two conventional schemes coincide when $M = 1$.

The value of P_m and P_{fa} represents the PU protection capability and spectrum opportunity utilization of a cognitive system, respectively. In Figure 3.7, we evaluate the performance of the ACSS compared with the conventional schemes with 100 SUs and $P_m = P_{fa} = 10^{-2}, 10^{-3}$ under the average SNR $\gamma = -10$ dB. The simulation result shows that the ACSS acquires the available channels faster than the conventional schemes under different system requirements of P_m and P_{fa}, and the performance of fast spectrum acquisition increases when P_m and P_{fa} are increasing.

Figure 3.8 shows the performance of the ACSS compared with the conventional schemes with the number of SUs $M = 100$, $P_m = P_{fa} = 10^{-2}$ under the average SNR $\gamma = -10$ dB, -15 dB. It is obvious that the ACSS acquires the available channels faster than the conventional schemes under different SNRs.

To further evaluate the performance of the schemes under different SNRs, we obtain the sensing time when 30 available channels are acquired, which is illustrated in Figure 3.9. In this figure, the horizontal axis represents the different SNRs, and the vertical axis represents the sensing time when 30 available channels are acquired. It is obvious that the ACSS outperforms the conventional schemes under all SNRs.

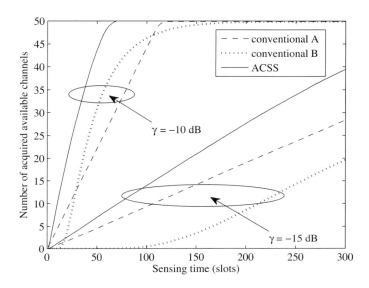

Figure 3.8: The performance of ACSS and conventional schemes with 100 SUs, $P_m = P_{fa} = 10^{-2}$ under different SNR averages.

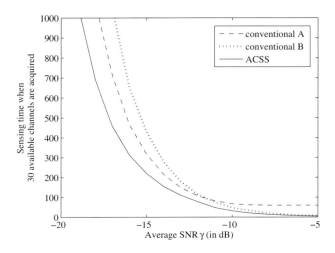

Figure 3.9: The average sensing time versus different SNRs, 30 available channels are acquired, $M = 100$.

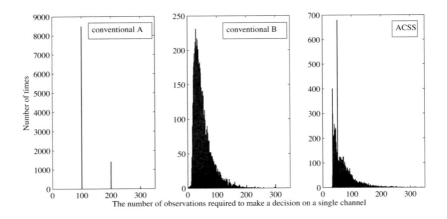

Figure 3.10: A histogram of the observation numbers before a decision on a single available channel is made.

In addition, when the SNR is higher than 9.0 dB, the sensing time of conventional scheme *A* does not reduce as the SNR increases. On the contrary, the sensing time of the ACSS and conventional scheme *B* reduces as the SNR increases. This is easy to understand. When the SNR is higher, no more than one idle channel will be acquired in one slot in conventional scheme *A*. However, more than one idle channel will be acquired in one slot in the ACSS and conventional scheme *B* since more than one channel is detected in one slot.

Furthermore, as shown in Figure 3.9, conventional scheme *A* outperforms conventional scheme *B* under low SNRs. On the contrary, conventional scheme *B* outperforms conventional scheme *A* under high SNRs. However, the ACSS scheme has the best performance under all SNRs. When the SNR is low, more SUs are expected to sense one certain channel at the same time for fast spectrum opportunity acquisition, therefore, conventional scheme *A* is more suitable for fast discovery than conventional scheme *B*. However, when the SNR is high, the channels are sensed and determined by fewer SUs, therefore, conventional scheme *B* is more suitable for fast discovery than conventional scheme *A* since more channels are sensed at the same time in conventional scheme *B*. Actually, the conventional schemes are two specific cases of the ACSS scheme. When the SNR is low, the ACSS will assign more SUs to sense one certain channel at the same time, which is similar to conventional scheme *A*. On the contrary, when the SNR is high, the ACSS will assign fewer SUs to sense one certain channel and more channels are sensed at the same time, which is similar to conventional scheme *B*. Hence, the ACSS is adaptive with respect to different SNRs.

Figure 3.10 shows a histogram of the number of observations before making a decision on a single available channel. In this simulation, 10^4 Monte Carlo searching has been done. $M = 100$, $P_m = P_{fa} = 10^{-2}$ and $\gamma = -10$ dB are considered. The horizontal axis represents the number of observations required to make a decision on

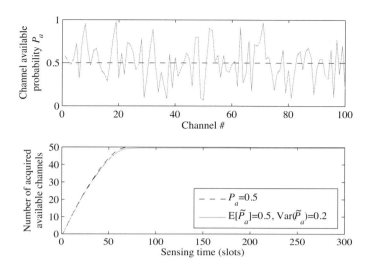

Figure 3.11: The robust performance of the ACSS with respect to inexactly P_a.

a single channel, and the vertical axis represents the number of times that certain observation numberd appeared in the simulation. As shown in Figure 3.10, the number of observations in conventional scheme A is largest and the number of observations in conventional scheme B is smallest. In conventional scheme A, the number of observations is $100 \times d$, $d = 1, 2, \cdots$, since $M = 100$ observations are obtained in each slot. Obviously, many observations are utilized to make a decision. In conventional scheme B, one observation is obtained in each slot, thus, all observations are necessary for making a decision. In the ACSS, the number of observations obtained in each slot is adaptive from 1 to 100. In fact, the ACSS makes a trade-off between the observation numbers and sensing time (i.e., the number of sensing slots).

In addition, since the probability of the channel being available, P_a and the average SNR γ are required to be known in the ACSS, we must estimate the value of P_a and γ when using the ACSS. However, the estimation may not be precise in practice. First, let $\tilde{P}_a$ and P_a denote the actual value and estimated value of the probability of the channel being available, respectively, $\tilde{\gamma}$ and γ denote the actual value and estimated value of the average SNR, respectively. In the following, we evaluate the robust performance of the ACSS via simulations. We assume $\tilde{P}_a$ follows a truncated normal distribution [59] with range $(0, 1)$, and the mean is equal to P_a (i.e., 0.5 in our simulation), the variance is equal to 0.2. As shown in Figure 3.11, the performance loss of fast spectrum acquisition is very small even though $\tilde{P}_a$ may be much different from P_a. Moreover, we assume that $\tilde{\gamma}$ follows a truncated normal distribution with range $(0, +\infty)$, and the mean is equal to γ (i.e., -10 dB in our simulation), the variance is equal to 0.01 and 0.03, respectively. As shown in Figure 3.12, the performance loss of fast spectrum acquisition is also small even though $\tilde{\gamma}$ may be much different from γ. Therefore, the ACSS is robust with respect to P_a and γ.

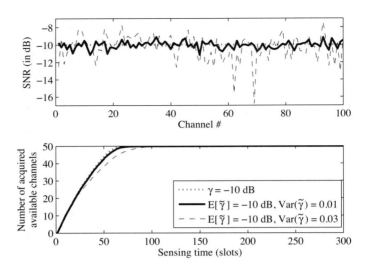

Figure 3.12: The robust performance of the ACSS with respect to inexact γ.

3.4 Summary

In this chapter, we have constructed a novel MAC-layer sensing framework for efficient acquisition of spectrum opportunities in a cognitive radio network with cooperative sensing. Specifically, we have employed the sequential probability ratio test and developed a new adaptive collaboration sensing scheme for multi-users to collaborate during multi-slots, in which the resources of secondary users are effectively utilized to sense the channels for efficient acquisition of spectrum opportunities. Subsequently, we have formulated the adaptive collaboration sensing scheme as an optimization problem and derived the optimal solution with low complexity based on dynamic programming theory. Simulation results show that the adaptive collaboration sensing scheme offers good performance with efficient acquisition of spectrum opportunities.

Appendix 3-A: Proof of $L\left(t_{ik}\right)$

Based on (3.2), the probability density function (PDF) of t_{ik} can be obtained by

$$f\left(t_{ik}|H_{i0}\right) = \frac{1}{\sqrt{2\pi\sigma_0^2}} \exp\left(\frac{-(t_{ik} - u_0)^2}{2\sigma_0^2}\right), \tag{3.15}$$

$$f\left(t_{ik}|H_{i1}\right) = \frac{1}{\sqrt{2\pi\sigma_1^2}} \exp\left(\frac{-(t_{ik} - u_1)^2}{2\sigma_1^2}\right), \tag{3.16}$$

where u_0 and σ_0^2 are the mean and variance of t_{ik} under H_{i0}, and $u_0 = \sigma_v^2$, $\sigma_0^2 = \frac{\sigma_v^4}{S}$, u_1 and σ_1^2 are the mean and variance of t_{ik} under H_{i1}, and $u_1 = \sigma_v^2(1 + \gamma_k)$, $\sigma_1^2 = \frac{\sigma_v^4}{S}(1 + 2\gamma_k)$.

Then, the LLR of t_{ik} is obtained by

$$
\begin{aligned}
L(t_{ik}) &= \ln \frac{f(t_{ik}|H_{i1})}{f(t_{ik}|H_{i0})} \\
&= \ln \frac{\sigma_0}{\sigma_1} + \frac{(t_{ik} - u_0)^2}{2\sigma_0^2} - \frac{(t_{ik} - u_1)^2}{2\sigma_1^2} \\
&= \ln \frac{\sigma_0}{\sigma_1} + \left(\frac{1}{2\sigma_0^2} - \frac{1}{2\sigma_1^2} \right) t_{ik}^2 \\
&\quad - \left(\frac{u_0}{\sigma_0^2} - \frac{u_1}{\sigma_1^2} \right) t_{ik} + \left(\frac{u_0^2}{2\sigma_0^2} - \frac{u_1^2}{2\sigma_1^2} \right) \\
&= A_k t_{ik}^2 - B_k t_{ik} + C_k \\
&= A_k \left(t_{ik} - \frac{B_k}{2A_k} \right)^2 - \frac{B_k^2}{4A_k} + C_k
\end{aligned}
\tag{3.17}
$$

where $A_k = \frac{1}{2\sigma_0^2} - \frac{1}{2\sigma_1^2} = \frac{S\gamma_k}{\sigma_v^4(1+2\gamma_k)}$, $B_k = \frac{u_0}{\sigma_0^2} - \frac{u_1}{\sigma_1^2} = \frac{S\gamma_k}{\sigma_v^2(1+2\gamma_k)}$, $C_k = \ln \frac{\sigma_0}{\sigma_1} + \left(\frac{u_0^2}{2\sigma_0^2} - \frac{u_1^2}{2\sigma_1^2} \right) = -\frac{1}{2}\ln(1 + 2\gamma_k) - \frac{S\gamma_k^2}{2(1+2\gamma_k)}$.

$A_k(t_{ik} - \frac{B_k}{2A_k})^2$ has a non-central chi-square distribution since t_{ik} has a normal distribution and A_k, B_k, C_k are constant for certain SUs. The mean and variance of the corresponding normal distribution are

$$
\begin{cases}
m_{0k} = \frac{\sqrt{S\gamma_k}}{2\sqrt{(1+2\gamma_k)}}, & \sigma_{0k}^2 = \frac{\gamma_k}{(1+2\gamma_k)}, & \text{under } H_{i0} \\
m_{1k} = \frac{\sqrt{S\gamma_k(1+2\gamma_k)}}{2}, & \sigma_{1k}^2 = \gamma_k, & \text{under } H_{i1}
\end{cases}
\tag{3.18}
$$

Hence, the proof is completed.

Appendix 3-B:
Derivation of $P\left(\sum\limits_{k=1}^{U_i} T_{ik} < \ln \eta_0 - \xi_i \right)$

Let F denote $P\left(\sum\limits_{k=1}^{U_i} T_{ik} < \ln \eta_0 - \xi_i \right)$. From (5.7), we can obtain

$$
F = P\left(\sum_{k=1}^{U_i} A_k \left(t_{ik} - \frac{B_k}{2A_k} \right)^2 < D_i \right),
\tag{3.19}
$$

where $D_i = \ln \eta_0 - \xi_i + \sum\limits_{k=1}^{U_i} \left(\frac{B_k^2}{4A_k} - C_k \right)$. Let X_k denote $A_k \left(t_{ik} - \frac{B_k}{2A_k} \right)^2$ and Y_i denote

$\sum\limits_{k=1}^{U_i} X_k$. From Appendix 3-A, we know X_k has a non-central chi-square distribution with a freedom of 1. Hence, we can easily obtain the characteristic functions of random variable Y_i when we assume the random variables X_k are statistically independent

$$
\begin{aligned}
\phi_{Y_i} &(jv) \\
&= E\left(e^{jvY_i} \right) \\
&= E\left[\exp\left(jv \sum_{k=1}^{U_i} X_k \right) \right] \\
&= \int_{-\infty}^{\infty} \cdots \int_{-\infty}^{\infty} (\prod_{k=1}^{U_i} e^{jvx_k}) p(x_1, \cdots x_{U_i}) \, dx_1 \cdots dx_{U_i} \\
&= \prod_{k=1}^{U_i} \phi_{X_k}(jv),
\end{aligned}
\tag{3.20}
$$

where $\phi_{X_k}(jv) = \left(1 - j2v\sigma_k^2 \right)^{-\frac{1}{2}} \exp\left(\frac{jvm_k^2}{(1-j2v\sigma_k^2)} \right)$ is the characteristic function of X_k, σ_k^2, and m_k^2 can be obtained by (5.10).

Then, the probability density function of Y_i is obtained by

$$
f_i(y) = \frac{1}{2\pi} \int_{-\infty}^{\infty} (\prod_{k=1}^{U_i} \phi_{X_k}(jv)) e^{-jvy} dv,
\tag{3.21}
$$

and F can be obtained by

$$
\begin{aligned}
F &= P\left(\sum_{k=1}^{U_i} A_k \left(t_{ik} - \frac{B_k}{2A_k} \right)^2 < D_i \right) \\
&= P(Y_i < D_i) \\
&= \int_{-\infty}^{D_i} f_i(y) \, dy \\
&= \frac{1}{2\pi} \int_{-\infty}^{D_i} \int_{-\infty}^{\infty} (\prod_{k=1}^{U_i} \phi_{X_k}(jv)) e^{-jvy} dv dy.
\end{aligned}
\tag{3.22}
$$

For a Rayleigh fading channel, γ_k has an exponential probability density function with the parameter γ, which is the average of γ_k. As illustrated in Figure 3.1, the distance among SUs are much smaller than the distance between SUs and PUs in our considered scenario. Hence, it is reasonable to assume that all SUs have the same average SNR γ [60]. However, it is still difficult to compute F in (3.22). We try to approximate F in (3.22) by using the average SNR γ instead of the SNR of the k-th SU γ_k, $k = 1, 2, \cdots, M$. Then, the approximate Y_i, denoted as $\bar{Y}_i$, has a non-central

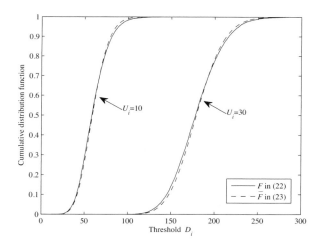

Figure 3.13: The cumulative distribution function of F and $\bar{F}$ with 100 SUs, $P_m = P_{fa} = 10^{-2}$.

chi-square distribution with freedom U_i, and $\bar{F}$ can be obtained by

$$\bar{F} = P\left(\sum_{k=1}^{U_i} \bar{A}_k \left(\bar{t}_{ik} - \frac{\bar{B}_k}{2\bar{A}_k}\right)^2 < \bar{D}_i\right) \tag{3.23}$$
$$= P(\bar{Y}_i < \bar{D}_i).$$

To evaluate the gap between F and $\bar{F}$, we obtain the cumulative distribution function (CDF) of F via Monte Carlo searching, and $\bar{F}$ via the CDF of a non-central chi-square distribution, respectively. In this simulation, we consider one idle channel and the average SNR $\gamma = -10$ dB. As shown in Figure 3.13, the CDFs of F and $\bar{F}$ match well under $U_i = 10, 30$, which illustrates that $\bar{F}$ approximates F well.

In addition, when the channel has not been determined, it is reasonable to assume that the channel is idle with probability P_a and busy with probability $1 - P_a$, where P_a is the probability of each channel being available. We also evaluate the performance loss if P_a cannot be estimated exactly in Section 3.3. Then, $\bar{F}$ can be obtained by

$$\bar{F} = P(\bar{Y}_i < \bar{D}_i)$$
$$= P_a P(\bar{Y}_i < \bar{D}_i | H_{i0}) + (1 - P_a)P(\bar{Y}_i < \bar{D}_i | H_{i1}), \tag{3.24}$$

where $P(\bar{Y}_i < \bar{D}_i | H_{i0})$ and $P(\bar{Y}_i < \bar{D}_i | H_{i1})$ are easily obtained by the CDF of a non-central chi-square distribution (the corresponding parameters of the non-central chi-square distribution can be obtained by (5.10)).

Hence, the expression of $P\left(\sum_{k=1}^{U_i} T_{ik} < \ln \eta_0 - \xi_i\right)$ in (3.9) is obtained.

Chapter 4

Transmission Power Allocation in a Cognitive Radio Network for Cellular Communication

Cooperative networking has received significant attention as an emerging network design strategy since future cellular networks are eager for higher capacity and larger coverage due to tremendously growing end-user demands and the amount of wireless terminals. One of the ways is to deploy more base stations (BSs) as done in traditional cellular networks. In contrast a cooperative relay-aided cellular network, as a more advanced system, introduces the use of relay stations (RSs) to increase the capacity and the coverage, which provides better quality of service (QoS) especially for cellular users (CUs) at the cell edge [61]. Moreover, cooperative relaying can also reduce overall costs compared with traditional non-relay approaches [62]. However, the performance gain of adopting cooperative relaying in cellular networks can be minor due to two major reasons: (1) a bottleneck link between the relay station and cellular user due to non-line-of-sight transmission, and (2) half-duplex cooperation that basically leads to a 50% throughput reduction of the transmission from the BS to the CU and may offset potentially achievable benefits from cooperative diversity, as depicted in Figure 1.7b.

In this chapter, a cognitive radio-assisted cooperation (CRAC) framework is introduced for the downlink transmissions in an orthogonal frequency division multiple access (OFDMA)-based cognitive radio network for cellular communications [64], where relay station leverages cognitive radio techniques to access white space sub-channels for relaying transmissions to CUs. Specifically, CRAC can support concurrent BS-to-RS/CU and RS-to-CU transmissions, and in turn overall downlink network throughput can be improved in contrast to traditional cooperative relaying. Each relay station is equipped with a cognitive radio and can opportunistically access sub-channels in the white space, that is, the frequency band that is not temporarily occupied by primary users [3]. Note that here primary users refer to the users in other networks operating in different frequency bands. With the additional white space sub-channels, full-duplex cooperation can be established. As illustrated in Figure 1.7b, the base station employs its dedicated band sub-channel C_1 for direct communication to the cellular user and relay station, while the relay station uses orthogonal sub-channel C_2 in the white space to relay transmissions to the cellular user. As a result, the BS-to-RS/CU and RS-to-CU transmissions can be delivered simultaneously. In turn, the overall achievable rate from the base station to the cellular user is improved. Moreover, if the white space lies in the ultra-high-frequency band (54–862 MHz) that has better propagation capability than the commonly used cellular band (e.g., 2000 MHz and higher frequency band), the RS-to-CU links have much better qualities to overcome the bottleneck effect. In short, such cognitive cooperative relaying in CRAC has the potential to greatly improve system performance. Moreover, a centralized optimization framework is introduced to maximize the network utility in the cognitive radio-assisted cooperation framework. Joint resource allocation, that includes transmission mode selection, relay station allocation, and transmit power/sub-channel allocation is considered to provide services and applications cost-effectively.

4.1   Cognitive Radio-Assisted Cooperation Framework

As shown in Figure 4.1, the CRAC framework considers an OFDMA-based cellular cell and some overlapping PU networks. The cell consists of a single BS, multiple CUs, and several RSs. RSs are deployed at some predefined locations inside the cell to aid BS-to-CU communications with cooperative relaying. In CRAC, each RS is equipped with two radios, which can transmit and receive data simultaneously over different sub-channels. Specifically, one radio uses dedicated band sub-channels (DBSs) and the other one as a cognitive radio uses white space sub-channels (WSSs). Moreover, the transmit power budget of RSs is constrained due to their low cost setup and/or limited energy source. Each PU network has an access point and some PUs. The cellular cell and PU networks operate over different frequency bands. For PU networks, CUs and RSs look like secondary users because they opportunistically access the band of PU networks using cognitive techniques. For the sake of clarity, we mainly consider the downlink transmission case in this chapter, and the CRAC framework can also be extended to the uplink transmission case. Without loss of generality and for simplicity, we make some assumptions as follows:

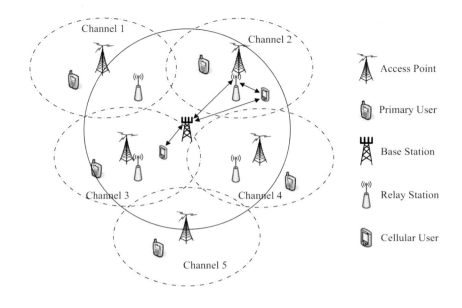

Figure 4.1: A cognitive radio-assisted cooperation framework.

(1) Cellular network

We assume that the RS adopts the decode-and-forward (DF) protocol with the same codebook as that of the BS. One RS may serve multiple CUs over different sub-channels simultaneously, and the downlink transmissions to a CU can occur over multiple DBSs. Moreover, we suppose that each DBS and WSS have the same bandwidth, and each transmission from the BS to a CU over one DBS can be assisted with one RS that relays the BS-to-CU transmission over one sub-channel (DBS in half-duplex cooperation or WSS in full-duplex cooperation). The frame level synchronization throughout the whole cellular network is assumed in this chapter. As the network operates in a slow fading environment, the accurate channel estimation is possible so that the full channel side information (CSI) is available at the BS, and the CSI stays static during an OFDMA frame. Moreover, a sub-channel can only be assigned to one CU, meaning that there is no sharing or intra-cell sub-channel reuse.

(2) PU network

We assume that some PU networks are overlapped with the cellular cell. Multiple RSs can perform independent or cooperative spectrum sensing to locate and exploit WSS while preventing the PUs from excessive interference [65]. Each PU network operates over an orthogonal licensed channel, which spans over some WSSs and has a certain protection area. We define that if the RS-to-CU link is within the protection

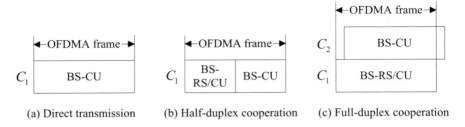

(a) Direct transmission   (b) Half-duplex cooperation   (c) Full-duplex cooperation

*For (c), the time offset between two copies of the data stream results from the DF delay.

Figure 4.2: An OFDMA frame structure for three transmission modes.

area of a PU network, it can utilize the PU channel only when the PU network is idle. Otherwise, the RS-to-CU link can always utilize the PU channel when its transmit power is not that high. Due to the geographical diversity, different RS-to-CU links may experience different WSS availabilities. This diversity can be well exploited by CRAC to produce more transmission opportunities. We suppose that the PU network state (active/idle) remains the same during an OFDMA frame.

We then introduce the operation of some elements in a cognitive radio-assisted cooperation framework.

(1) BS operations

With the purpose of reducing the cost of updating infrastructures, the BS does not have any functionality with the cognitive radio, and it solely uses the DBSs. In the CRAC framework, there is a built-in centralized resource scheduler at the BS, which allocates resources to cater for downlink data streams of all CUs in the cell at the beginning of each OFDMA frame. The resource scheduler jointly considers the selection of transmission modes, the assignment of sub-channels, RSs, and transmit power, with the knowledge of the CSI of BS-to-RS, BS-to-CU, and RS-to-CU links over all sub-channels obtained during the current OFDMA frame. The scheme of obtaining an optimal resource allocation policy will be discussed in the following.

(2) RS operations

As stated above, each RS is equipped with two radios that can transmit and receive data over DBSs and WSSs simultaneously. During the spectrum sensing duration at the beginning of each OFDMA frame, the RS performs spectrum sensing and channel probing to obtain the CSI of RS-to-CU links over WSS and DBS. After that, the RS operates under three transmission modes according to the scheduling of the BS.

As depicted in Figure 4.2, there are three transmission modes for downlink data streams and accordingly three OFDMA frame structures.

- **Direct transmission.** The RS stays idle over the DBS C_1 during the current OFDMA frame when the data stream is over a certain DBS C_1.

- **Half-duplex cooperation.** When the data stream is over a certain DBS C_1, the RS assists the data stream transmission over the same DBS C_1. In the first half-frame, the RS receives and decodes the data stream from the BS. In the second half-frame, the RS encodes and forwards the received data stream to the corresponding CU, while the BS stays idle over the DBS C_1.

- **Full-duplex cooperation.** Similarly, the RS assists the data stream transmission over an additional WSS C_2. Therefore, the RS decodes and forwards the data stream symbol by symbol over the WSS C_2, while the BS still occupies the DBS C_1.

(3) CU operations

The CU can receive data streams over DBSs and WSSs simultaneously with two independent radios working over two different frequency bands. As shown in Figure 4.2, when the data stream over one DBS is assisted by a RS, the CU combines the two copies of the data stream into two time slots (see Figure 4.2b) or over two sub-channels (see Figure 4.2c) with maximum ratio combining [66].

In summary, the CRAC framework aims to improve the network throughput through efficient scheduling of resources (transmit power, sub-channels, RSs, and so forth), which will be addressed in detail in the following.

Denote Φ, Ω, ψ, and ξ as the set of CUs, RSs, DBSs, and WSSs, respectively. $m \in \Phi = \{1, 2, \cdots, K\}$ represents the index of a CU as well as the index of the combined data stream[1] received at a CU. $r \in \Omega = \{1, 2, \cdots, J\}$ represents a RS. $c_D \in \psi = \{1, 2, \cdots, N\}$ and $c_W \in \xi = \{N+1, N+2, ..., N+SG\}$ represent a DBS and a WSS, respectively. Moreover, suppose that S PU networks coexist with the cellular cell, and each PU channel spans over G WSSs ($K, J, N, S, G \in \mathbf{N}_+$).

Denote $\mathbf{Q}(t) = \{Q_s(t)\}_S$ as the state of the PU network during the OFDMA frame t, where $Q_s(t)$ is a binary value. $Q_s(t) = 1$ means that the PU network over the channel s is active. Otherwise, $Q_s(t) = 0$. According to a finite state ergodic Markov chain, $\mathbf{Q}(t)$ evolves. Depending on the current locations of the RS and the CU, as well as the PU network state, a RS-to-CU link can access a subset of PU channels potentially during an OFDMA frame. Since each RS can relay data streams for all CUs in the cell, the maximum amount of possible RS-to-CU links is JK. Therefore, we define the PU channel availability matrix $\mathbf{V}(t) = \{V_{r,m}^s(t)\}_{S \times JK}$ as

$$V_{r,m}^s(t) = \begin{cases} 0, \text{when the link of } r\text{th RS to} \\ \quad m\text{th CU over channel } s \text{ is available,} \\ 1, \text{otherwise.} \end{cases} \tag{4.1}$$

---

[1] Here, a combined data stream is defined as a combination of all data streams received at the CU from the BS and RSs over multiple sub-channels.

As mentioned above, each data stream operates in one and only one of three transmission modes over one DBS as shown in Figure 4.2. For simplicity, the noises are modeled as independent and identically distributed (i.i.d.) circularly symmetric complex Gaussian noise $\mathscr{CN}(0, N_0 W)$, where W is the bandwidth of a DBS or a WSS, and N_0 is the power density of the noises.

(1) Direct transmission mode

For convenience, we first define $R_1(m, c_D, P_{BS}^{c_D})$ as the achievable rate function over the DBS $c_D \in \psi$, which is allocated to the combined data stream m, where $P_{BS}^{c_D}$ denotes the transmit power at the BS over the DBS c_D. With the spectral efficiency formulation (in b/s/Hz) [67], we have

$$R_1(m, c_D, P_{BS}^{c_D}) = \log_2 \left(1 + \frac{P_{BS}^{c_D} \cdot |h_{m,c_D}|^2}{\Gamma N_0 W} \right), \tag{4.2}$$

where Γ is the gap to Shannon capacity, which is mainly determined by modulation techniques and the target bit-error rate, and h_{m,c_D} denotes the channel gain from the BS to the mth CU over the DBS c_D. For simplicity, Γ is set to one in this chapter.

(2) Half-duplex cooperation mode

Similarly, we define $R_2(m, c_D, P_{BS}^{c_D}, r, P_r^{c_D})$ as the achievable rate function over the DBS $c_D \in \psi$ that is allocated to the combined data stream m, where RS $r \in \Omega$ is assigned to be the relay over the DBS c_D with the allocated power $P_r^{c_D}$. In the half-duplex cooperation mode, the BS and RS transmit over the same DBS in a two time-slots sharing manner. Therefore, according to Laneman, Tse, and Wornell [61], the achievable rate is

$$\begin{aligned} &R(m, c_D, P_{BS}^{c_D}, r, P_r^{c_D}) \\ &= \frac{1}{2} \min \left\{ \log_2 \left(1 + \frac{2 P_{BS}^{c_D} \cdot |h_{r,c_D}|^2}{\Gamma N_0 W} \right), \ \log_2 \left(1 + \frac{2 P_{BS}^{c_D} \cdot |h_{m,c_D}|^2 + 2 P_r^{c_D} \cdot |h_{r,m,c_D}|^2}{\Gamma N_0 W} \right) \right\}, \end{aligned} \tag{4.3}$$

where the coefficient $1/2$ before the right part of the formulation and the coefficient 2 before $P_{BS}^{c_D}$ and $P_r^{c_D}$ account for the fact that the half-duplex cooperation operates over two time slots, respectively. h_{r,c_D}, h_{m,c_D}, and h_{r,m,c_D} denote the channel gain from the BS to the rth RS, BS to the mth CU, and rth RS to the mth CU over the DBS c_D, respectively.

(3) Full-duplex cooperation mode

Similarly, we define $R_3(m, c_D, P_{BS}^{c_D}, r, c_W, P_r^{c_W})$ as the achievable rate function over the DBS $c_D \in \psi$ and the WSS $c_W \in \xi$ that are allocated to the combined data stream m. RS $r \in \Omega$ is assigned to be the relay with the allocated power $P_r^{c_W}$ over the WSS $c_W \in \xi$. In the full-duplex cooperation mode, the BS and RS transmit over different sub-channels in parallel. Therefore, the achievable rate is

$$
\begin{aligned}
&R(m, c_D, P_{BS}^{c_D}, r, c_W, P_r^{c_W}) \\
&= \min\left\{ \log_2\left(1 + \frac{P_{BS}^{c_D} \cdot |h_{r,c_D}|^2}{\Gamma N_0 W}\right), \log_2\left(1 + \frac{P_{BS}^{c_D} \cdot |h_{m,c_D}|^2 + P_r^{c_W} \cdot |h_{r,m,c_W}|^2}{\Gamma N_0 W}\right)\right\},
\end{aligned}
$$
(4.4)

where h_{r,c_D} and h_{m,c_D} denote the channel gain from the BS to the rth RS and BS to the mth CU over the DBS c_D, respectively. h_{r,m,c_W} is the channel gain from the rth RS to the mth CU over the WSS c_W. Since the direct data transmission and the relay data transmission are conducted simultaneously in the long run with the full-duplex cooperation mode, (4.4) does not have the coefficient of $1/2$.

To find the proper trade-off between the network throughput and the fairness among CUs, the sum utility maximization framework is employed as the objective function in this chapter. Obviously, the utility function is a concave and increasing function of the achievable rate for the data stream of each CU and reflects the satisfaction of the CU.

For convenience, we define $U_m(\cdot)$ as the utility function of achievable rate of the combined data stream m. Moreover, three 0–1 indicators $\alpha_m^{c_D}$, β_m^{r,c_D}, and γ_m^{r,c_D,c_W} are defined, in which $\alpha_m^{c_D}$ indicates whether combined data stream m is assigned DBS c_D for the direct transmission or not, β_m^{r,c_D} denotes whether combined data stream m is operating in the half-duplex cooperation mode with rth RS and DBS c_D, or not, and γ_m^{r,c_D,c_W} represents whether combined data stream m is operating in the full-duplex cooperation mode with rth RS, DBS c_D, and WSS c_W, or not. Thus, the achievable rate of the combined data stream m can be expressed as

$$
\begin{aligned}
\lambda_m = &\sum_{c_D \in \psi} R_1(m, c_D, P_{BS}^{c_D})\alpha_m^{c_D} + \sum_{c_D \in \psi}\sum_{r \in \Omega} R_2(m, c_D, P_{BS}^{c_D}, r, P_r^{c_D})\beta_m^{r,c_D} \\
&+ \sum_{c_D \in \psi}\sum_{c_W \in \xi}\sum_{r \in \Omega} R_3(m, c_D, P_{BS}^{c_D}, r, c_W, P_r^{c_W})\gamma_m^{r,c_D,c_W},
\end{aligned}
$$
(4.5)

where $\forall m \in \Phi$.

Therefore, the optimization objective function is

$$
\max \sum_{m \in \Phi} U_m(\lambda_m),
$$
(4.6)

constraints are given as follows.

(1) DBS allocation constraints

As mentioned above, each DBS can only be assigned to one data stream which operates with one of the three modes,

$$
\sum_{m \in \Phi}\left(\alpha_m^{c_D} + \sum_{r \in \Omega}\beta_m^{r,c_D} + \sum_{c_W \in \xi}\sum_{r \in \Omega}\gamma_m^{r,c_D,c_W}\right) \leq 1, \forall c_D \in \psi.
$$
(4.7)

(2) WSS allocation constraints

Obviously, each WSS can be assigned to assist only one DBS,

$$\sum_{m\in\Phi}\sum_{c_D\in\psi}\sum_{r\in\Omega}\gamma_m^{r,c_D,c_W}\le 1,\forall c_W\in\xi. \tag{4.8}$$

Moreover, each assigned WSS should be available to its corresponding RS-to-CU link. Define mapping function $s=f(c_W)$ as that WSS c_W is spanned over by PU channel s, we have

$$\sum_{m\in\Phi}\sum_{c_D\in\psi}\sum_{r\in\Omega}\left[\gamma_m^{r,c_D,c_W}\cdot V_{r,m}^{f(c_W)}\right]=0,\forall c_W\in\xi. \tag{4.9}$$

(3) Power constraints

$$\sum_{c_D\in\psi}P_{BS}^{c_D}\le\bar{P}_{BS}, \tag{4.10}$$

$$\sum_{c_D\in\psi}P_r^{c_D}+\sum_{c_W\in\xi}P_r^{c_W}\le\bar{P}_{RS},\forall r\in\Omega. \tag{4.11}$$

4.2 Optimal Transmission Power Allocation

Obviously, the optimization problem of (4.6) is a mixed integer programming problem. According to Yu and Lui [70], the optimization problem of (4.6) can be treated as a spectrum balancing problem in an OFDMA system with a large number of subchannels, which can be solved optimally via convex optimization techniques. Therefore, the Lagrangian dual decomposition method is adopted to obtain the optimal solution of (4.6).

4.2.1 Cross-Layer Optimization

For the ease of utilizing the Lagrangian dual decomposition, we introduce an auxiliary value, that is, the application layer demand of each CU denoted by d_m ($\forall m\in\Phi$). We then rewrite (4.6) as

$$\max\sum_{m\in\Phi}U_m(d_m), \tag{4.12}$$

s.t. (4.7)–(4.11), and $\forall m$, $\lambda_m\ge d_m$.

Obviously, the objective of (4.12) is maximized when $\lambda_m=d_m$ $\forall m$ because U_m is a concave and increasing function. As a result, (4.6) and (4.12) have the same solution. Introducing Lagrange multiplier vector $\boldsymbol{\theta}$, the dual function can be written as

$$g(\boldsymbol{\theta})=\begin{cases}\max\limits_{m\in\Phi}\sum[U_m(d_m)+\theta_m(\lambda_m-d_m)],\\ \text{s.t. }(4.7)-(4.11),\end{cases} \tag{4.13}$$

where θ_m ($m \in \Phi$) is the element of $\boldsymbol{\theta}$. The dual function (4.13) can be separated into two maximization sub-problems. The first sub-problem is a utility maximization problem, corresponding to a rate adaptation problem in the application layer,

$$g_{appl}(\boldsymbol{\theta}) = \max_{d_m, \forall m} \sum_{m \in \Phi} [U_m(d_m) - \theta_m d_m]. \tag{4.14}$$

The second problem is a joint RS, transmit power, sub-channel allocation, and transmission mode selection problem in the physical layer,

$$g_{phy}(\boldsymbol{\theta}) = \begin{cases} \max \sum_{m \in \Phi} \theta_m \lambda_m, \\ \text{s.t. } (4.7) - (4.11). \end{cases} \tag{4.15}$$

Due to the fact that the sum utility maximization problem of (4.12) has a zero duality gap when the number of sub-channels go to infinity [70], it can be solved by minimizing the dual objective as

$$\begin{aligned} &\min g(\boldsymbol{\theta}), \\ &\text{s.t. } \boldsymbol{\theta} \succeq \mathbf{0}. \end{aligned} \tag{4.16}$$

According to Boyd, Xiao, and Mutapcic [71], the problem of (4.16) can be solved with updating $\boldsymbol{\theta}$ via the sub-gradient method. Note that the dual variable $\boldsymbol{\theta}$ can be considered as the price of the achievable rate. When the CU demand exceeds the achievable rate, θ_m grows. Otherwise, θ_m decreases.

Compared with the sub-problem (4.15), it is easier to solve the sub-problem (4.14). According to Ng and Yu [69], we define the utility function of each combined data stream as

$$U(x) = \begin{cases} k_1 \left(1 - e^{-k_2 x}\right), \text{if } x \geq 0, \\ -\infty, \qquad\quad \text{if } x < 0, \end{cases} \tag{4.17}$$

where k_1 determines the upper bound of the utility function. k_2 reflects the average demand level of the CU and it should be carefully tuned according to application layer requirements and fairness rules.

4.2.2 Power Constraint Elimination

In this section, we give a solution of (4.15). For simplicity, we assume that the transmit power over each DBS at the BS is equal and predefined as P_{BS}^o and $P_{BS}^o \leq \frac{\bar{P}_{BS}}{N}$. Then, a Lagrange multiplier vector $\boldsymbol{\mu}$ is introduced in the power constraint (4.11) and the dual function is written as

$$a(\boldsymbol{\mu}) = \begin{cases} \max \sum_{m \in \Phi} \theta_m \lambda_m + \sum_{r \in \Omega} \mu_r \left(\bar{P}_r - \sum_{c_D \in \psi} P_r^{c_D} - \sum_{c_W \in \xi} P_r^{c_W} \right) \\ \text{s.t.} (4.7), (4.8). \end{cases} \tag{4.18}$$

where μ_r ($r \in \Omega$) is the element of $\boldsymbol{\mu}$.

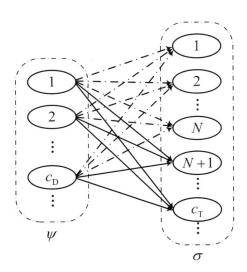

Figure 4.3: An equivalent CRAC problem and weighted bipartite matching. Not all links are shown here.

Since $\sum_{r\in\Omega}\mu_r\bar{P}_r$ is constant in (4.18) when μ_r is given, we rewrite (4.18) as

$$\max \sum_{m\in\Phi}\theta_m\lambda_m - \sum_{r\in\Omega}\left(\mu_r\sum_{c_D\in\psi}P_r^{c_D}\right) - \sum_{r\in\Omega}\left(\mu_r\sum_{c_W\in\xi}P_r^{c_W}\right),$$
$$\text{s.t.} (4.7),(4.8),$$

(4.19)

which is termed as the CRAC problem. Obviously, the CRAC problem (4.19) is equivalent to the dual problem (4.18).

Since (4.18) has a zero duality gap [70], it can be solved by minimizing the dual objective function as

$$\min a(\boldsymbol{\mu}),$$
$$\text{s.t.} \boldsymbol{\mu} \succeq \mathbf{0},$$

(4.20)

where we remove the constraint (4.9) by setting the channel gain to zero when the link over a WSS is not available due to the PU activities. Similarly, we solve this dual problem via updating $\boldsymbol{\mu}$ with the sub-gradient method, where the dual variable $\boldsymbol{\mu}$ can be considered as the price of relay power.

4.2.3 Throughput Maximization

As depicted in Figure 4.3, we transform the CRAC problem to an equivalent weighted bipartite matching problem. We then give the following proposition.

Proposition 4.1

The optimal solution of the equivalent weighted bipartite matching problem is also the optimal solution of the CRAC problem.

Proof: For convenience, we construct a bipartite graph $A = (\psi \times \sigma, E)$, where ψ is the set of DBS, σ is the combined set of DBS and WSS, $\sigma = \psi \cup \xi$. Moreover, c_T denotes a DBS when $c_T \in [1, N]$ and denotes a WSS when $c_T \in [N + 1, N + SG]$. The edge set E corresponds to $|\psi||\sigma|$ edges connecting all possible pairs of sub-channels in two vertex sets. Each edge (c_D, c_T) carries four attributes, $(w_{c_D,c_T}, m_{c_D,c_T}, r_{c_D,c_T}, P_{c_D,c_T})$, where weight w_{c_D,c_W} is mapped from the effective achievable rate of the data stream m_{c_D,c_T} assisted by the RS r_{c_D,c_T} with transmit power P_{c_D,c_T} over sub-channel set $\{c_D, c_T\}$. Note that the effective achievable rate accounts for the price of the achievable rate and relay power, that is, θ and μ. Through this mapping, the maximum matching process represents the comprehensive assignment of RSs, sub-channels, and transmit power, as well as the transmission mode selection. Moreover, constraints (4.7) and (4.8) in CRAC problem (4.19) are also inherent in the matching process. As a result, the optimal solution of the weighted bipartite matching problem is also the optimal solution of the CRAC problem. □

In the following, we obtain four attributes in three cases.

Case 1: $c_D = c_T \in [1, N]$. In this case, the transmission mode can be direct transmission or half-duplex cooperation (dotted links in Figure 4.3). Define the effective achievable rate functions as

$$g_1(m) = \theta_m R_1(m, c_D, P_{BS}^o), \tag{4.21}$$

$$g_2(m, r, P_r^{c_D}) = \theta_m R_2(m, c_D, P_{BS}^o, r, P_r^{c_D}) - \mu_r P_r^{c_D}, \tag{4.22}$$

then, we have

$$w_{c_D,c_T} = \max \left\{ \max_m g_1(m), \max_{m,r,P_r^{c_D}} g_2(m, r, P_r^{c_D}) \right\}. \tag{4.23}$$

To maximize $g_1(m)$, search $m \in \Phi$ and pick out the optimal $\hat{m}$. Moreover, the optimal $(\hat{m}, \hat{r}, \hat{P}_r^{c_D})$ is picked out to maximize $g_2(m, r, P_r^{c_D})$.

When $\max_m g_1(m) \geq \max_{m,r,P_r^{c_D}} g_2(m, r, P_r^{c_D})$, direct transmission mode is better for the data stream than the half-duplex cooperation mode. Thus, we have

$$m_{c_D,c_T} = \arg\max_m g_1(m), \tag{4.24}$$

and $r_{c_D,c_T} = 0$, $P_{c_D,c_T} = 0$. Otherwise, we have

$$(m_{c_D,c_T}, r_{c_D,c_T}, P_{c_D,c_T}) = \arg\max_{m,r,P_r^{c_D}} g_2(m, r, P_r^{c_D}). \tag{4.25}$$

Case 2: $c_D \neq c_T \in [1, N]$ (dotted links in Figure 4.3). Obviously, it is impossible to match between two different DBSs in this case. Therefore, we set w_{c_D,c_T}, m_{c_D,c_T}, r_{c_D,c_T}, and P_{c_D,c_T} to zero.

Case 3: $c_T \in [N + 1, N + SG]$. Obviously, the transmission mode is full-duplex cooperation with DBS c_D and WSS c_T (the solid links in Figure 4.3). Define the effective achievable rate function

$$g_3(m, r, P_r^{c_T}) = \theta_m R_3(m, c_D, P_{BS}^o, r, c_T, P_r^{c_T}) - \mu_r P_r^{c_T}, \qquad (4.26)$$

then, we have

$$w_{c_D, c_T} = \max_{m, r, P_r^{c_T}} g_3(m, r, P_r^{c_T}), \qquad (4.27)$$

and

$$(m_{c_D, c_T}, r_{c_D, c_T}, P_{c_D, c_T}) = \arg \max_{m, r, P_r^{c_T}} g_3(m, r, P_r^{c_T}). \qquad (4.28)$$

The optimal $(\hat{m}, \hat{r}, \hat{P}_r^{c_T})$ is picked out to maximize $g_3(m, r, P_r^{c_T})$.

So far, some good polynomial-time algorithms have been introduced to optimally solve the bipartite matching problem, for example, the Hungarian algorithm [72]. Obviously, the whole algorithm is in polynomial time since the computation complexity of the graph construction is $O(|\psi| |\sigma| |\Phi| |\Omega|)$.

4.3 Performance Analysis and Evaluation

4.3.1 Simulation Scenario

In conducted computer simulations, a cell is centered at the BS with a radius of 1000 meters. $J = 4$ fixed RSs are located at $(\pm 353.5, \pm 353.5)$, that is, on a circle with the radius equal to 500 meters. CUs are uniformly distributed inside the cell. The channels between the BS, RSs, and CUs are selected as the COST-231 model [68], including the path loss, large-scale shadowing and small-scale fading. Suppose that the BS and RSs are both placed at some height above the ground, thus, the fading has a line-of-sight (LOS) component. The cell area is overlapped with $S = 3$ PU networks. The protection radius of each PU network is set to 1000 meters. Each PU network operates over one PU channel that spans over $G = 3$ OFDMA sub-channels. The state of each PU channel is modeled as a Markov ON-OFF process. The bandwidth of a sub-channel (DBS or WSS) is $W = 200$ kHz. There are $N = 30$ DBSs centered around 2000 MHz, while WSSs are centered around 700 MHz. The transmit power constraints of BS and RS are 25 dBm and 19 dBm, respectively. The white noise power density is -174 dBm/Hz. The parameters of the utility function are determined via setting the satisfactory throughput of each CU to 7.0 Mbps. The satisfactory throughput achieves a utility value that is 90% of the maximum achievable utility. For comparison, we consider three different schemes as follows.

■ **BL-1:** It is the baseline scheme that only uses a direct transmission mode without cooperative relaying.

■ **BL-2:** It is the baseline scheme that uses both the direct transmission mode and half-duplex cooperation mode (i.e. traditional cooperative relaying).

■ **CRAC:** It is the optimal scheme in the CRAC framework.

The following performance metrics are considered and calculated in the simulation for the comparison of performance between the CRAC scheme and baseline schemes as described above.

■ **Network utility:** It is defined as the sum of K CUs' utilities. The network utility reflects the satisfaction of the CUs, which ranges between 0 and 1.

■ **Network throughput:** It is defined as the sum of K CUs' throughput in the cell.

■ **Throughput gain:** It is defined as the ratio of the network throughput of a scheme to that of the BL-1 scheme. If the gain of a scheme considered is bigger than 100%, it means that the scheme achieves better performance than the BL-1 scheme.

■ **Fairness:** Jain's fairness index [73] is adopted as the fairness metric, which is defined as

$$f = \frac{\left(\sum_{m=1}^{K} \lambda_m\right)^2}{K \sum_{m=1}^{K} \lambda_m^2} \qquad (4.29)$$

where λ_m is the achievable rate of the mth CU. The Jain's fairness index ranges between $1/K$ and 1.0. When it approaches 1.0, the system is fairer.

4.3.2 Performance Comparisons

In this section, the number of CUs is set to $K = 4$. The probabilities of the cases that a PU network transforms from ON (active) to OFF (idle) and *vice versa* are 0.2, and 0.2, respectively. Therefore, the PU traffic load is $\rho = 0.5$. Performance comparisons among different schemes are depicted in Figure 4.4. In summary, we have

■ Cooperative relaying schemes (including BL-2 and CRAC) can achieve a higher network utility than the non-cooperative scheme (e.g., the BL-1). Moreover, the network utility of the CRAC scheme is the highest among all three schemes.

■ The network throughput is moderately improved by the BL-2 scheme with traditional cooperative relaying up to 10% compared with the BL-1 scheme. However, the CRAC scheme can greatly improve the network throughput up to 76.5% even when the PU traffic load is heavy ($\rho = 0.5$) due to its capability of exploiting white space with optimal resource allocation.

■ Better fairness can be obtained with the BL-2 scheme, compared with the BL-1 scheme. However, the CRAC scheme achieves the best fairness among all the schemes.

Obviously, the performance of the CRAC scheme significantly outperforms the other two baseline schemes.

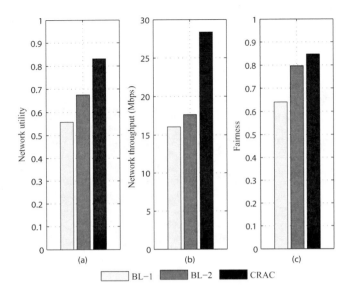

Figure 4.4: Performance with three different schemes.

4.3.3 Impact of the Cell Population

As shown in Figure 4.5, it is obvious that the network utilities of the three schemes decrease as the cell population grows. The major reason is that each CU has less resources on average as the number of CUs become larger. Similarly, the network throughput of the three schemes increases as the cell population climbs due to user diversity gain, which is shown in Figure 4.6. In addition, Figure 4.7 shows that the fairness of the three schemes becomes worse as the cell population goes up. However, the CRAC scheme still achieves much better performance in network utility, network throughput, and fairness than the other two baseline schemes with different cell populations.

4.3.4 Impact of the Primary User Traffic Load

Figure 4.8 shows some curves of throughput gain versus PU traffic load with the CRAC and the BL-2 schemes under three different numbers of CUs. It is obvious that the throughput gain of the CRAC scheme decreases as the PU traffic load varies from 0 (lightest) to 1 (heaviest) with different numbers of CUs. When the PU traffic load is 0, the resulting throughput gain is the upper-bound performance of the CRAC scheme. When the number of CUs is $K = 4$, we find that the upper bound of the throughput gain is only slightly higher than that of the PU traffic load $\rho = 0.5$ by 6.25%. It is also obvious that the CRAC scheme still has a larger throughput gain than that of the BL-2 scheme when the PU traffic load $\rho = 1$. All results show the robustness of the CRAC scheme to the change of the PU traffic load, which is because

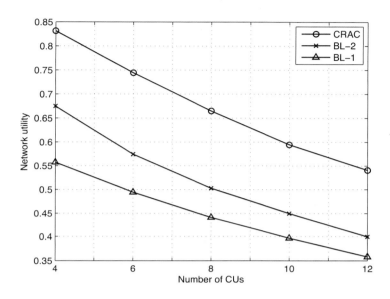

Figure 4.5: Network utility versus the number of CUs with three different schemes.

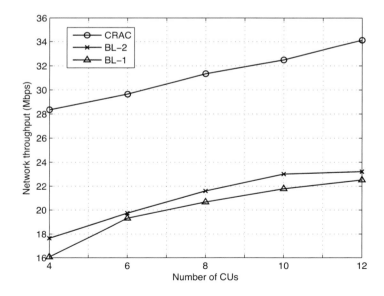

Figure 4.6: Network throughput versus the number of CUs with three different schemes.

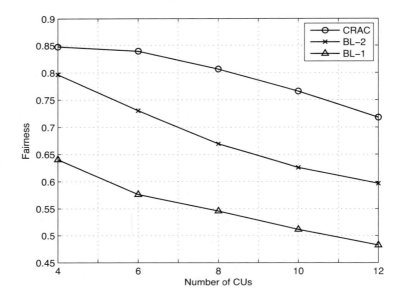

Figure 4.7: Fairness versus the number of CUs with three different schemes.

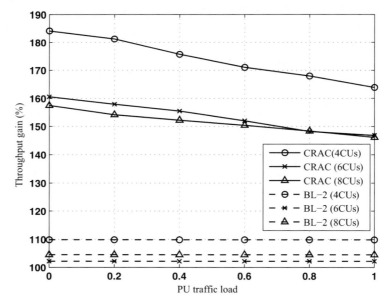

Figure 4.8: Throughput gain versus PU traffic load with the CRAC and BL-2 schemes with three different numbers of CUs.

the spatial diversity offers more abundant transmission opportunities in the cognitive radio environment [74]. Moreover, we find that the throughput gain of the CRAC scheme outperforms that of the BL-2 scheme under different numbers of CUs with the varying of the PU traffic loads.

4.4 Summary

In this chapter, we introduced a novel CRAC framework for downlink transmissions in an OFDMA-based cognitive radio network for cellular communications. A centralized optimal solution via employing the dual decomposition method was derived for the CRAC framework. Extensive simulation results show a significant performance gain of the CRAC framework when compared with traditional cooperative relaying in terms of the network utility, network throughput, and fairness.

Chapter 5

White Space Allocation in a Cognitive Radio-Based High-Speed Vehicle Network

It is well known that high-speed vehicles, such as high-speed trains, are playing an increasingly important role in people's lives, since they provide a relatively stable and spacious environment for long distance travelers. As a result, there is a strong demand for broadband wireless communications (BWC) for high-speed vehicles to provide information access and onboard entertainment services for passengers. However, the cognitive radio network should be employed to overcome the problem of spectrum scarcity.

For a cognitive radio network, primary channel occupancy modeling is very important for designing secondary user spectrum accessing and scheduling schemes. PUs in different primary networks usually have different activity models, such as in a cellular network [76], wireless local area network (WLAN) [27], TV/broadcasting network [77], and so forth. However, the existing literature is concerned about static or low-speed mobile users, and do not consider situations with high-speed mobile users, in which, the channel availabilities are very different due to the high velocity of users [78, 79].

In this chapter, high-speed vehicles are considered for employing the TV/broadcasting band via cognitive radio technology to achieve broadband wireless communications [106]. In a high-speed moving scenario, primary channel occupancy is mainly dependent on the secondary user's position, moving speed, and direction. Therefore, the relative primary channel occupancy model, which mainly depends on the vehicle's location, velocity, moving direction, and so on, is considered. Moreover, a cognitive radio-based high-speed vehicle network (CR-HSVN) is introduced and a framework of the spectrum resource allocation is introduced for effectively utilizing TV white space.

5.1   A Cognitive Radio-Based High-Speed Vehicle Network

5.1.1   System Model

In this chapter, we employ the following terms: (1) primary base station (PBS): the TV signal transmission tower; (2) primary user (PU): the TV set; (3) cognitive base station (CBS): the CBSs are deployed in the considered area similar to the base stations in the cellular network, and are managed and operated by wireless operators; (4) cognitive vehicle (CV): high-speed vehicle; and (5) cognitive user (CU): the passenger in the CV that needs to access the cognitive radio network. The main variables are listed in Table 5.1.

As shown in Figure 5.1, we assume that there are L PBSs in a certain area, denoted as $\mathscr{P}_l, l = 1, 2, \cdots, L$. N primary channels, denoted as $\mathscr{C}_n, n = 1, 2, \cdots, N$, are assigned to PBSs to provide services for PUs in their coverage ranges. Each PBS $\mathscr{P}_l$ occupies a fixed and known primary channel $\mathscr{C}_n$. Note that when $N \leqslant L$, different PBSs may use the same primary channel if they are far away from each other. Locations of these PBSs are known. The coverage radius r_l of $\mathscr{P}_l$ may be different for different PBSs according to the actual demand. The PBSs with overlapped

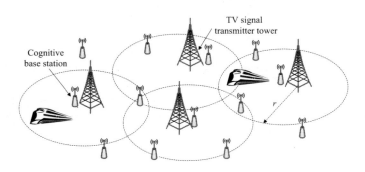

Figure 5.1: The scenario of CR-based high-speed vehicle communications.

Table 5.1 List of the Main Variables

L	the number of PBSs
N	the number of primary channels
M	the number of vehicles
$\psi_1^{\mathscr{P}_l}$	the communication range of the PBS $\mathscr{P}_l$
$\psi_2^{\mathscr{P}_l}$	the protection range of the PBS $\mathscr{P}_l$
ϕ_1^m	the communication range of the m-th vehicle
ϕ_2^m	the interference (to other vehicles) range of the m-th vehicle
ϕ_3^m	the interference (to PUs) range of the m-th vehicle
$d_{m,\mathscr{P}_l}$	the distance between the m-th vehicle and the PBS $\mathscr{P}_l$
$d_{m,\tilde{m}}$	the distance between the m-th vehicle and the $\tilde{m}$-th vehicle
d_{th}	the distance threshold to constrain the aggregated interference
$P_{m,n}$	the availability indicator of the n-th primary channel for the m-th vehicle
$F_{m,\tilde{m},n}$	the availability indicator between $\mathscr{S}_m$ and $\mathscr{S}_{\tilde{m}}$ on $\mathscr{C}_n$
A_m	the channel requirement of the m-th vehicle
$\delta_{m,n}$	the channel assignment indicator of the n-th primary channel for the m-th vehicle
R_1, R_2, R_3	spectrum utilizations in the three levels
J	the number of the minimum independent vehicle sets
Q	the number of the rows of the relationship matrix (**RM**)
w_m	the number of rows that include the element m in the relationship matrix (**RM**)
u_t	the index number of vehicles that satisfies $F_{m,u_t} = 1$ or $F_{u_t,m} = 1$, $t = 1, 2, \cdots, w_m$
$\mathbf{B}$	the set that consists of the ID of $\delta_{u,n}$ if $\delta_{u,n} = 1$
q	the element number of the set $\mathbf{B}$
$\boldsymbol{\delta}^*$	the optimal solution to (5.15)
$\boldsymbol{\delta}^\circ$	the optimal solution to (5.17)
I	the maximum aggregated interference in the CR-HSVN
G	Jain's fairness index
x_m	the available spectrum resource allocated to the m-th vehicle

coverage areas occupy different primary channels to avoid interference. Obviously, these assumptions are reasonable since the information of the PBSs can easily be obtained from the TV and spectrum management agencies. Denote $\mathscr{S}_m$, $m = 1, 2, \cdots, M$, as the m-th CV which passes through the areas covered by TV signals. Then, the CV employs cognitive radio technology to access the idle primary channels for communications with the CBS.

Obviously, the state of TV channels is nearly stable for static or low-speed vehicles when the PBS is always active or the duration of busy/idle is long. However, for high-speed vehicles, the busy duration of certain channels may be short, since high-speed vehicles quickly pass through the coverage area of certain TV towers. In

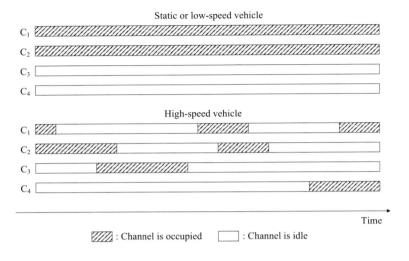

Figure 5.2: Relative primary channel occupancy model.

this chapter, the TV signal activity model is illustrated in Figure 5.2 for high-speed vehicles, which is called the relative primary channel occupancy model.

The availability of licensed channels for the CBS can be obtained based on the Geolocation Database. Different CBSs can exchange information with each other and manage CVs to access TV channels. One CV may contain one or multiple CUs, and CUs in the same CV have the same spectrum availability. As shown in Figure 5.3, for

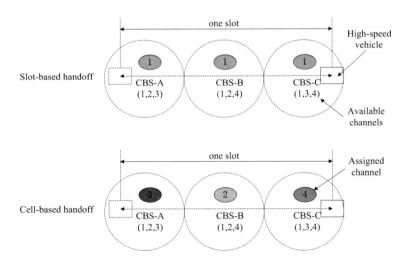

Figure 5.3: Slot-based spectrum handoff.

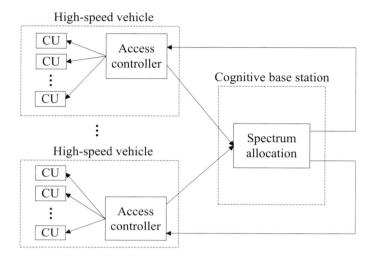

Figure 5.4: The system model of the CR-HSVN.

high-speed movement, the vehicle may cross multiple CBSs in a slot. This will lead to the challenge that the spectrum handoff is done many times in one slot because the availability of channels at each CBS can be different. To avoid frequent spectrum handoffs for high-speed CVs, a slot-based handoff method is employed in this chapter. In the slot-based handoff method, only the channels that are available at all the CBSs that a vehicle may cross in one slot can be utilized. For example, as shown in Figure 5.3, the vehicle would cross CBS-A, CBS-B, and CBS-C in one slot, and each CBS would have different available channels. By employing our proposed slot-based handoff method in the system, the vehicle can only use channel 1 for wireless service in this slot. Note that CBSs can allocate channels to the CVs at the beginning of each slot, and the duration of the slot can be set according to the velocity and moving direction of the vehicle.

The CR-HSVN system model is illustrated in Figure 5.4, in which there are multiple CUs and an access controller in each CV. The access controller sends the location, velocity, moving direction, and channel requirement to the CBSs via existing vehicular systems. The CBSs collect the information to calculate the list of available channels and the list of spectrum sharing, respectively. Then, the CBSs obtain the optimal spectrum allocation strategy to maximize the spectrum utilization of the CR-HSVN. Note that the spectrum utilization denotes the number of channels used by CUs over the number of whole channels in a given area. Finally, the allocation strategy can be transmitted to the access controller in each CV, and the access controller can control the CUs to access the assigned channels.

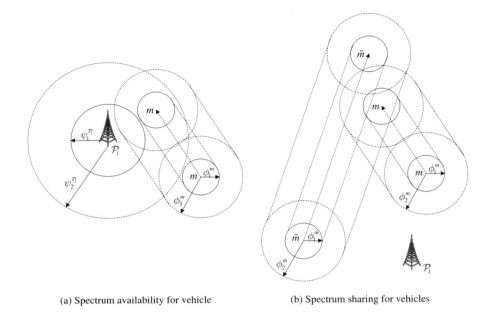

(a) Spectrum availability for vehicle (b) Spectrum sharing for vehicles

---

Figure 5.5: Analysis of ACL and SSL.

5.1.2 Path Loss Model

As shown in Figure 5.5, we introduce some ranges as follows:

- **TV communication range** $\psi_1^{\mathscr{P}_l}$: This is the effective coverage range of the PBS $\mathscr{P}_l$, in which the PUs can effectively receive TV signals from the PBS $\mathscr{P}_l$.

- **TV protection range** $\psi_2^{\mathscr{P}_l}$: The CVs in this range cannot use the same frequency with the corresponding PBS $\mathscr{P}_l$. Usually, $\psi_2^{\mathscr{P}_l} > \psi_1^{\mathscr{P}_l}$. Note that the fading margin has been included in this range.

- **Vehicle communication range** ϕ_1^m: The m-th CVs can correctly communicate with the corresponding CBS in this range.

- **Vehicle interference (to other vehicles) range** ϕ_2^m: This is the coverage range of the m-th vehicle, where the interference to other vehicles is higher than certain thresholds. Usually, $\phi_2^m > \phi_1^m$.

- **Vehicle interference (to PUs) range** ϕ_3^m: This is the coverage range of the m-th vehicle, where the interference to PUs is higher than certain thresholds. Usually, $\phi_3^m > \phi_2^m > \phi_1^m$.

Table 5.2   Coefficients of the Model

a_{00}	a_{01}	a_{02}	a_{03}	a_{04}
2.6191	0.0318005	-9.50112×10^{-4}	1.46844×10^{-5}	-8.30291×10^{-8}
a_{10}	a_{11}	a_{12}	a_{13}	a_{14}
-3.63991×10^{-3}	2.4824×10^{-4}	-1.15328×10^{-5}	1.98061×10^{-7}	-1.06459×10^{-9}
a_{20}	a_{21}	a_{22}	a_{23}	a_{24}
7.20911×10^{-6}	-6.73582×10^{-7}	3.19009×10^{-8}	-5.48948×10^{-10}	2.96093×10^{-12}
a_{30}	a_{31}	a_{32}	a_{33}	a_{34}
-5.75331×10^{-9}	5.86862×10^{-10}	-2.80726×10^{-11}	4.83956×10^{-13}	-2.61299×10^{-15}
a_{40}	a_{41}	a_{42}	a_{43}	a_{44}
1.48675×10^{-12}	-1.5699×10^{-13}	7.53695×10^{-15}	-1.29985×10^{-16}	7.01903×10^{-19}

These ranges are determined when the corresponding SNR, transmitted power, and path loss are given.

For TV signals, the path loss can be modeled as Perez-Vega and Zamanillo [83]

$$L_{TV}^{s}(d_1) = 10k \lg(d_1) + L_{TV}^{0}, \qquad (5.1)$$

where L_{TV}^{s} (in dB) is the path loss between the PBS and PU, d_1 (in meter) is the distance between the PBS and PU, L_{TV}^{0} (in dB) is the attenuation at 1.0 meter in free space, that is,

$$L_{TV}^{0} = 20 \lg \left(\frac{4\pi}{\lambda} \right), \qquad (5.2)$$

where λ is the wavelength. The investigation results show that the parameter k is independent of frequency [84]. Moreover, based on the FCC F(50,50) propagation curves for television broadcasting in the VHF and UHF bands [85], the values of k for different distances and transmitting antenna heights were obtained in Perez-Vega and Zamanillo [83], and several attempts to fit the value of k to mathematical functions were done, such as, logarithmic, exponential, and polynomial. The best fit was obtained with a polynomial model of the fourth degree as

$$k = \sum_{i=0}^{4} \sum_{j=0}^{4} a_{ij} h^i d_1^j, \qquad (5.3)$$

where h is the transmitting antenna height, d_1 is the distance between the PBS and PU.

In Perez-Vega and Zamanillo [83], a fitting was performed with Stanford Graphics software, which can provide good data analysis and create presentation-oriented documents, to obtain the coefficients a_{ij} when h is in feet and d_1 is in miles. These coefficients are given in Table 5.2. Then, the TV communication range ψ_1 and protection range ψ_2 can be determined by Equations (5.1), (5.2), and (5.3) when the received SNR threshold of the PU and the sensing SNR threshold (which is set to be -20 dB in the IEEE 802.22 standard) are given.

Then, as the TV tower transmitted power is given in Table 5.3, and the received SNR threshold of the PU and the sensing SNR threshold (which is set to be -20

dB in the IEEE 802.22 standard) are given, the TV communication range $\psi_1^{\mathscr{P}_l}$ and protection range $\psi_2^{\mathscr{P}_l}$ can be calculated by Equations (5.1), (5.2), and (5.3).

For high-speed vehicles, we employ the multi-path propagation model as the path loss model. Currently, various mean path loss models have been developed, and one of the popular models is the Okumura-Hata propagation prediction model [87], in which the mean path loss is expressed as

$$
\begin{aligned}
L_{\text{train}}^{\text{s}}(d_2) = {} & 69.55 + 26.16 \lg(f) - 13.82 \lg(h_{\text{b}}) - a(h_{\text{m}}) \\
& + [44.9 - 6.55 \lg(h_{\text{b}})] \lg(d_2) - L_{\text{rur}}^{\text{s}},
\end{aligned}
\tag{5.4}
$$

where $L_{\text{train}}^{\text{s}}$ (in dB) is the path loss between train and CBS, f (in MHz) is the operation frequency, h_{m} (in meters) is the height of the train's antenna, h_{b} (in meters) is the height of the CBS's antenna, and d_2 (in kilometers) is the distance between the vehicle and CBS. $L_{\text{rur}}^{\text{s}}$ (in dB) is the correction factor when trains are in a rural area, and it can be written as

$$
L_{\text{rur}}^{\text{s}} = 4.78 \left(\lg(f)\right)^2 - 18.33 \lg(f) + K,
\tag{5.5}
$$

where K is related to the geographical environment and chang from 35.94 in the countryside to 40.94 in the desert.

Since we employ the slot-based handoff method in the system, a CV may cross several CBSs in one slot. The vehicle communication range ϕ_1^m can be obtained according to the maximum distance between the m-th vehicle and the nearest CBS in each slot. Then, the transmitted power of the vehicles is determined. Note that the transmitted power of each CV is constant in the same slot but can be varied in different slots due to the different vehicle communication range ϕ_1^m. The vehicle interference range ϕ_2^m and ϕ_3^m can be determined by Equations (5.4) and (5.5) when the thresholds of the interference to the PUs and CVs are given.

5.1.3 Available Channel List

The available channel list (ACL) is a $M \times N$ binary matrix with element $P_{m,n}$, where $P_{m,n} = 1$ indicates that the vehicle $\mathscr{S}_m$ can use $\mathscr{C}_n$ for transmission, and $P_{m,n} = 0$ indicates that the vehicle $\mathscr{S}_m$ cannot use $\mathscr{C}_n$ for transmission. For low-speed or static CVs, whether the primary channels are available or not, it is only related to the location of the CVs. Hence, the ACL is nearly a constant matrix. However, for high-speed cognitive vehicles, the primary channel occupancy model is closely related to some parameters of the CVs, such as location, velocity, and moving direction. As shown in Figure 5.5a, we assume that $\mathscr{C}_n$ is assigned to the PBS set $\mathscr{P}_{\text{set}}^n = \{\mathscr{P}_{n1}, \mathscr{P}_{n2}, \cdots\}$, then, the ACL can be obtained as

$$
P_{m,n} = \begin{cases} 1, & \psi_1^{\mathscr{P}_l} + \phi_3^m < d_{m,\mathscr{P}_l} \text{ and } \psi_2^{\mathscr{P}_l} < d_{m,\mathscr{P}_l} \text{ in the current slot for } \mathscr{P}_l \in \mathscr{P}_{\text{set}}^n, \\ 0, & \text{else}, \end{cases}
\tag{5.6}
$$

where $d_{m,\mathscr{P}_l}$ is the distance between the PBS $\mathscr{P}_l$ and the vehicle $\mathscr{S}_m$. The condition $\psi_1^{\mathscr{P}_l} + \phi_3^m < d_{m,\mathscr{P}_l}$ indicates that the PUs in the coverage range of $\mathscr{P}_l$ are not located in the interference range of the m-th vehicle ϕ_3^m. The condition $\psi_2^{\mathscr{P}_l} < d_{m,\mathscr{P}_l}$ indicates that the m-th vehicles are not located in the protection range of the PBS $\mathscr{P}_l$.

5.1.4 Spectrum Sharing List

The spectrum sharing list (SSL) is an $M \times M \times N$ three-dimensional binary matrix with element $F_{m,\tilde{m},n}$, where $m,\tilde{m} = 1,2,\cdots,M, n = 1,2,\cdots,N$. The SSL illustrates whether it will cause harmful interference or not when two vehicles share a certain channel in the current slot. First, when vehicle $\mathscr{S}_m$ and $\mathscr{S}_{\tilde{m}}$ are very close, they cannot share $\mathscr{C}_n$ to avoid harmful interference to each other. In addition, it is necessary to consider the aggregated interference in the spectrum sharing system. Since the received power is proportional to $d^{-\alpha}$ when the transmission power is fixed, where d is the distance between the transmitter and receiver, α is the path loss exponent, $\alpha = 2 \sim 4$ and $\alpha = 2$ only when free space is considered. Therefore, the interference with the smallest d plays a major role in the aggregated interference. In this chapter, to constrain the aggregated interference, we first constrain the major interference, and then let the vehicles that share the same channel have different distances with PU [1] using the corresponding channel.

Then, let $F_{m,\tilde{m},n} = 1$ indicate that the vehicles $\mathscr{S}_m$ and $\mathscr{S}_{\tilde{m}}$ cannot share $\mathscr{C}_n$, and let $F_{m,\tilde{m},n} = 0$ indicate that the vehicles $\mathscr{S}_m$ and $\mathscr{S}_{\tilde{m}}$ can share $\mathscr{C}_n$. As shown in Figure 5.5b, the SSL can be obtained as

$$F_{m,\tilde{m},n} = \begin{cases} 1, & \phi_1^{\tilde{m}} + \phi_2^m > d_{m,\tilde{m}} \text{ or } \phi_1^m + \phi_2^{\tilde{m}} > d_{m,\tilde{m}} \text{ or } |d_{m,\mathscr{P}_l} - d_{\tilde{m},\mathscr{P}_l}| < d_{th} \\ & \text{in the current slot for } \mathscr{P}_l \in \mathscr{P}_{\text{set}}^n \text{ and } \tilde{m} \neq m, \\ 0, & \text{else,} \end{cases}$$
(5.7)

where $d_{m,\tilde{m}}$ is the distance between the vehicles m and $\tilde{m}$. Obviously, $F_{m,\tilde{m},n} = F_{\tilde{m},m,n}$. $\phi_1^{\tilde{m}} + \phi_2^m > d_{m,\tilde{m}}$ or $\phi_1^m + \phi_2^{\tilde{m}} > d_{m,\tilde{m}}$ indicates that the vehicles $\mathscr{S}_m$ and $\mathscr{S}_{\tilde{m}}$ are too close, and as a result are not sharing the same channel. $|d_{m,\mathscr{P}_l} - d_{\tilde{m},\mathscr{P}_l}| < d_{th}$ indicates that the distances from $\mathscr{S}_m$ and $\mathscr{S}_{\tilde{m}}$ to $\mathscr{P}_l$ are close, hence, $\mathscr{S}_m$ and $\mathscr{S}_{\tilde{m}}$ should not share the channel with $\mathscr{P}_l$ to avoid large aggregated interference. d_{th} is the distance threshold and $d_{th} \in [0, +\infty)$. Note that a larger d_{th} means a stricter constraint to aggregated interference.

5.2 Maximization of Utilized White Space

Suppose that CVs have different service requirements with respect to spectrum resources, which is described by channel requirement A_m. In this chapter, we aim to maximize the spectrum utilization of the CR-HSVN. Consider that if the assigned spectrum resource to the CV $\mathscr{S}_m$ is larger than A_m, some spectrum resources will be wasted, which is contradictory to the goal. Therefore, the amount of channels

---

[1]For a simplified analysis, the PU is replaced by the PBS. In fact, this simplification reinforces the constraint of aggregated interference.

assigned to $\mathscr{S}_m$ should not be larger than A_m. The spectrum utilization is divided into three levels as follows:

- **Level 1:** All CVs can use the whole spectrum resources without any constraint such as interference to PUs and other CVs. Then, the spectrum resources in this level, denoted as R_1, can be written as

$$R_1 = \sum_{m=1}^{M} T_{1,m}, \qquad (5.8)$$

where $T_{1,m} = \min\{N, A_m\}$. Obviously, R_1 can be considered as an upper bound of the spectrum resources that can be utilized by the CR-HSVN.

- **Level 2:** Due to the activity of PUs, some spectrum resources in Level 1 are not used. The spectrum utilization in this level, denoted as R_2, can be written as

$$R_2 = \sum_{m=1}^{M} T_{2,m}, \qquad (5.9)$$

where $T_{2,m} = \min\{N, A_m, \sum_{n=1}^{N} P_{m,n}\}$.

- **Level 3:** To constrain the interference among vehicles and the aggregated interference to PUs, some spectrum resources in Level 2 are not used. The spectrum utilization in this level, denoted as R_3, can be written as

$$R_3 = \sum_{n=1}^{N} \sum_{m=1}^{M} \delta_{m,n} P_{m,n}, \qquad (5.10)$$

where $\delta_{m,n}$ is an indicator function and it can be written as

$$\delta_{m,n} = \begin{cases} 1, & \text{channel } \mathscr{C}_n \text{ is assigned to vehicle } \mathscr{S}_m, \\ 0, & \text{else.} \end{cases} \qquad (5.11)$$

Meanwhile, due to the activity of the PUs, $\delta_{m,n}$ must satisfy

$$\delta_{m,n}(1 - P_{m,n}) = 0. \qquad (5.12)$$

Moreover, to constrain the interference among vehicles and the aggregated interference to the PUs

$$\delta_{m,n} + \delta_{\tilde{m},n} \leqslant 2 - F_{m,\tilde{m},n}. \qquad (5.13)$$

To achieve the channel requirements of the vehicles, we have

$$\sum_{n=1}^{N} \delta_{m,n} \leqslant A_m. \qquad (5.14)$$

In this chapter, we aim to minimize the spectrum utilization loss due to interference to the PUs and to other CVs in the CR-HSVN, that is, minimize $|R_1 - R_3|$. Note that R_1 is independent on the allocation scheme and $R_1 \geqslant R_3$. Therefore, we only need to maximize R_3, that is, the optimization problem can be formulated as

$$\max_{\boldsymbol{\delta}} R_3$$

$$s.t. \begin{cases} \delta_{m,n}(1 - P_{m,n}) = 0, \\ \delta_{m,n} + \delta_{\tilde{m},n} + F_{m,\tilde{m},n} \leqslant 2, \\ \sum_{n=1}^{N} \delta_{m,n} \leqslant A_m, \\ \delta_{m,n} \in \{0,1\}, \end{cases} \tag{5.15}$$

where $\boldsymbol{\delta} = \{\delta_{m,n} | m = 1,2,\cdots,M; n = 1,2,\cdots,N\}$.

Obviously, (5.15) is a $0 - 1$ integer programming problem with a linear constraint (IPLC), and the number of variables is MN. The optimal solution can be obtained with exponential time complexity when 2^{MN} possible solutions are searched. Specifically, when the IPLC can be converted into an assignment problem, there are several standard algorithms that can be used to obtain the optimal solution with low complexity [88, 89]. However, the problem (5.15) cannot be converted into the assignment problem. For example, one vehicle can obtain multiple channels, meanwhile, one channel can also be assigned to multiple vehicles (the number of vehicles is unknown). Fortunately, under some special conditions, we can reduce the complexity to obtain the optimal solutions. In the following section, we will introduce the property of *variable independence* that makes it possible to obtain the optimal solution by separation computing, resulting in the reduction of the complexity. We then introduce the optimal solution based on an efficient branch and bound searching method, suboptimal solutions based on a single channel and linear programming with lower complexity, respectively.

5.2.1   Separation Computing

Note that the vehicles in the CR-HSVN can be divided into one or more vehicle sets, and each vehicle only belongs to one set. Here, we first define the *independent vehicle set*.

Definition 5.1

A vehicle set, denoted as an **V**, is considered an *independent vehicle set* if and only if any vehicle that belongs to **V** does not cause interference to any other vehicle that does not belong to **V**.

According to the definition, it is obvious that **V** is an independent vehicle set when **V** contains no vehicle (**V** $= \varnothing$) or **V** contains all vehicles (**V** $= I$). Moreover, if the number of vehicles in the independent vehicle set **V** equals 1, we call **V** the minimum independent vehicle set. Besides, if the number of vehicles in the independent

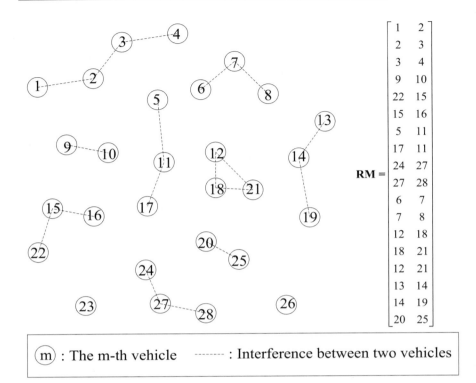

$$RM = \begin{bmatrix} 1 & 2 \\ 2 & 3 \\ 3 & 4 \\ 9 & 10 \\ 22 & 15 \\ 15 & 16 \\ 5 & 11 \\ 17 & 11 \\ 24 & 27 \\ 27 & 28 \\ 6 & 7 \\ 7 & 8 \\ 12 & 18 \\ 18 & 21 \\ 12 & 21 \\ 13 & 14 \\ 14 & 19 \\ 20 & 25 \end{bmatrix}$$

(m) : The m-th vehicle ⎯⎯⎯ : Interference between two vehicles

Figure 5.6: An example of independent vehicle sets.

vehicle set **V** is larger than 1, and for any vehicle belonging to **V**, there is at least one other vehicle that can cause interference to this vehicle, we also call **V** the minimum independent vehicle set. Note that the element number of the minimum independent vehicle set can be varied in different slots due to the movement of each vehicle. As an example, Figure 5.6 shows independent vehicle sets, where 28 vehicles are divided into 11 minimum independent vehicle sets.

We then consider finding the optimal solution to **V**, which means that we try to assign an appropriate spectrum resource to the vehicles in **V** to maximize the spectrum utilization of the vehicles in **V**. Obviously, finding the optimal solution to **V** = *I* is equal to finding the optimal solution to (5.15). Based on the definition and analysis, we have the following property.

Property 5.1 *Suppose vehicles in the CR-HSVN can be divided into J minimum independent vehicle sets, denoted as* $\mathbf{V}_j$, $j = 1, 2, \cdots, J$. *We can then obtain the optimal solution to* $\mathbf{V} = I$ *by separately obtaining the optimal solution to* $\mathbf{V}_j$.

Proof: Suppose the optimal solution to $\mathbf{V}_j$ is $\boldsymbol{\delta}_j^*$ and its corresponding optimal spectrum utilization is $R_{3,j}^*$, $j = 1, 2, \cdots, J$, which means that for any other solution to

$\mathbf{V}_j$, denoted as $\boldsymbol{\delta}_j$, the corresponding spectrum utilization is $R_{3,j} \leqslant R_{3,j}^*$. Then, the optimal solution of $\mathbf{V} = I$ is $\boldsymbol{\delta}^* = \{\boldsymbol{\delta}_1^* \cup \boldsymbol{\delta}_2^* \cup \cdots \cup \boldsymbol{\delta}_J^*\}$, and its corresponding optimal spectrum utilization is $R_3^* = \sum_{j=1}^{J} R_{3,j}^*$.

When $\boldsymbol{\delta}^*$ is not the optimal solution to $\mathbf{V} = I$, which means that there is another solution $\boldsymbol{\delta} = \{\boldsymbol{\delta}_1 \cup \boldsymbol{\delta}_2 \cup \cdots \cup \boldsymbol{\delta}_J\}$ that $\boldsymbol{\delta} \neq \boldsymbol{\delta}^*$, and its corresponding spectrum utilization is $R_3 = \sum_{j=1}^{J} R_{3,j}$, and $R_3 > R_3^*$. Without loss of generality, we divided $\boldsymbol{\delta}$ into two sets $\mathbf{X}$ and $\mathbf{Y}$, where $\boldsymbol{\delta}_j = \boldsymbol{\delta}_j^*$ if $j \in \mathbf{X}$ and $\boldsymbol{\delta}_j \neq \boldsymbol{\delta}_j^*$ if $j \in \mathbf{Y}$. Due to the independence of $\mathbf{V}_j$, the solution to $\mathbf{V}_j$ does not interfere with the solutions to $\mathbf{V}_i$ for $\forall i \neq j$. Therefore, $R_{3,j} = R_{3,j}^*$ if $j \in \mathbf{X}$ and $R_{3,j} \leqslant R_{3,j}^*$ if $j \in \mathbf{Y}$. Thus, under the solution $\boldsymbol{\delta}$, the corresponding spectrum utilization is $R_3 = \sum_{j \in \mathbf{X}} R_{3,j} + \sum_{j \in \mathbf{Y}} R_{3,j} \leqslant \sum_{j \in \mathbf{X}} R_{3,j}^* + \sum_{j \in \mathbf{Y}} R_{3,j}^* = R_3^*$, which is contradictory to $R_3 > R_3^*$. Thus, R_3^* is the optimal solution to $\mathbf{V} = I$.

This completes the proof of Property 5.1. □

According to Property 5.1, we can divide a problem with a large number of variables into some sub-problems with a small number of variables, thus resulting in a reduction of complexity. Moreover, we can solve the sub-problems in a separation method, that is, we can solve the sub-problems in different CBSs, which means parallel computing is possible and the efficiency of the system is greatly improved. For example, if $M = 50$, $N = 10$, the complexity of the problem is $O(2^{MN})$. However, if the 50 vehicles are divided into $J = 25$ minimum independent vehicle sets, and each set includes 2 vehicles, then the overall complexity is $O(J \times 2^{MN/J})$. Obviously, the smaller the element number of the minimum independent vehicle set the lower the overall complexity is.

Figure 5.7 is a flow chart that shows the algorithm for obtaining minimum independent vehicle sets for the CR-HSVN. According to the definition of the minimum independent vehicle set, the CVs $\mathscr{S}_m$ and $\mathscr{S}_{\tilde{m}}$ belong to the same set when $F_{m,\tilde{m},n} = 1$. We define the relationship matrix (**RM**) with Q rows and 2 columns. First, the element of **RM** denotes the index number of vehicles, that is, m or $\tilde{m}$. Second, m and $\tilde{m}$ are the two elements of a certain row of **RM** if and only if $F_{m,\tilde{m},n} = 1$. Therefore, Q is the number of vehicle pairs in **RM**. Figure 5.6 shows an example of independent vehicle sets and the corresponding **RM** with $Q = 18$. Third, let w_m denote the number of rows that include the element m, and $w_m = \sum_{\tilde{m}=1}^{M} \sum_{n=1}^{N} F_{m,\tilde{m},n}$. Let u_t denote the index number of CVs that satisfy $F_{m,u_t,n} = 1$ or $F_{u_t,m,n} = 1$, $t = 1, 2, \cdots, w_m$. Let ξ_m denote a minimum independent vehicle set. Then, based on the discussion above, we can obtain the minimum independent vehicle sets according to the flow chart described in Figure 5.7.

5.2.2 Branch and Bound Search Method

Let U denote the element number of a certain minimum independent vehicle set $\mathbf{V}^U$. First for $\mathbf{V}^U$, we assign a number ranging from 1 to UN for each variable as the address (ID) of the variable, that is, the ID of variable $\delta_{u,n}$ is considered as $(u-1)N + n$, where $u = 1, 2, \cdots, U$, $n = 1, 2, \cdots, N$. We then define set **B** that consists of the ID of $\delta_{u,n}$ if $\delta_{u,n} = 1$. For example, when $U = N = 5$, $\mathbf{B} = \varnothing$ denotes

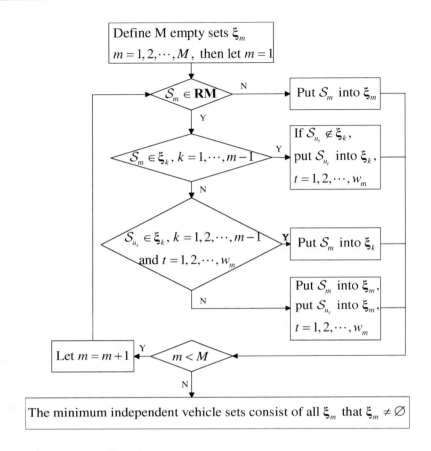

Figure 5.7: Flow chart to obtain the minimum independent vehicle sets.

that all the variables are equal to 0, and $\mathbf{B} = \{1, 8, 15\}$ denotes that $\delta_{1,1} = 1$, $\delta_{2,3} = 1$, $\delta_{3,5} = 1$, and the other variables are equal to 0. Obviously, each $\mathbf{B}$ is a possible solution to $\mathbf{V}^U$.

Let q denote the element number of $\mathbf{B}$, $q = 0, 1, \cdots, UN$. Then, the number of the sets with q elements is C_{UN}^q. Therefore, the total number of sets is $\sum_{q=0}^{UN} C_{UN}^q = 2^{UN}$, that is, the optimal solution to $\mathbf{V}^U$ can be obtained by searching all the sets. We then construct a combination tree according to the following rules:

(1) the nodes of the tree are made up of all possible $\mathbf{B}$;

(2) $\mathbf{B} = \varnothing$ is the root node of the combination tree;

(3) the elements of each node is in ascending order;

(4) each child node consists of the elements of its father node, and each child node has one more element than its father node. Moreover, the element that does not belong to the father node is larger than all the elements of the father node.

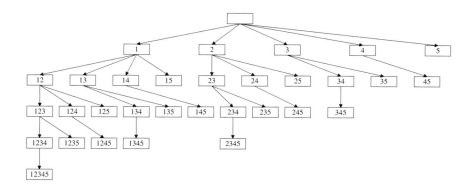

Figure 5.8: An example of a combination tree when the number of variables is 5.

Figure 5.8 shows an example of the combination tree when the number of variables is 5. Based on the combination tree, there is an important corollary as shown below.

Corollary 5.1

When a node does not satisfy the constraints in (5.15) (replace M with U), all children of the node will also not satisfy the constraint. As a result, all the nodes in the sub-tree with this node as the root will not satisfy the constraint.

The proof of Corollary 5.1 is omitted since it is a well-known property. According to Corollary 5.1, the optimal solution based on branch and bound searching has a lower complexity than exhaust searching in many cases, since the former method can reduce the number of searches according to the constraint in many cases. However, in the worst case, the number of searches based on branch and bound searching approaches the number of searches based on exhaust searching. Therefore, it is necessary to introduce some suboptimal solutions with a much lower complexity.

5.2.3 Single-Channel Method with Low Complexity

To reduce the complexity, we separately assign each channel to approximate the vehicles. For the channel $\mathscr{C}_n$, the following problem is solved

$$\max_{\boldsymbol{\delta}} \sum_{m=1}^{M} \delta_{m,n} P_{m,n}$$

$$s.t. \begin{cases} \delta_{m,n}\left(1 - P_{m,n}\right) \leqslant 0, \\ \delta_{m,n} + \delta_{\tilde{m},n} + F_{m,\tilde{m},n} \leqslant 2, \\ \delta_{m,n} \in \{0,1\}. \end{cases} \tag{5.16}$$

Then, for $\check{n} = 1, 2, \cdots, N$, we assign $\mathscr{C}_{\check{n}}$ to the vehicle $\mathscr{S}_m$ if $\delta_m^{\check{n}} = 1$ and $\sum_{n=1}^{\check{n}} \delta_m^n < A_m$.

5.2.4 Linear Programming Method with Low Complexity

Second, we relax the constraint $\delta_{m,n} \in \{0,1\}$ in (5.15) to $\delta_{m,n} \in [0,1]$. Then, the problem can be transformed as

$$\max_{\delta} R_3$$

$$s.t. \begin{cases} \delta_{m,n}(1-P_{m,n}) \leqslant 0, \\ \delta_{m,n} + \delta_{\tilde{m},n} + F_{m,\tilde{m},n} \leqslant 2, \\ \sum_{n=1}^{N} \delta_{m,n} \leqslant A_m, \\ \delta_{m,n} \in [0,1]. \end{cases} \tag{5.17}$$

Obviously, (5.17) is a linear programming problem and can be solved with some effective methods, such as a simplex algorithm. To analyze the condition of the non-integer solutions and truncate the non-integer solutions to 0 or 1, we investigate the Karush-Kuhn-Tucker (KKT) conditions of (5.17). We define

$$f(\delta_{m,n}) = \sum_{m,n} \delta_{m,n} P_{m,n} - \sum_{m,n} \mu_1^{n,m} \delta_{m,n}(1-P_{m,n})$$

$$- \sum_{m,\tilde{m},n} \mu_2^{n,m,\tilde{m}} \left(\delta_{m,n} + \delta_{\tilde{m},n} + F_{m,\tilde{m},n} - 2\right) - \sum_m \mu_3^m \left(\sum_n \delta_{m,n} - A_m\right)$$

$$- \sum_{m,n} \mu_4^{n,m}(\delta_{m,n} - 1) + \sum_{m,n} \mu_5^{n,m} \delta_{m,n}, \tag{5.18}$$

where the KKT multipliers $\mu_1^{n,m} \geqslant 0$, $\mu_2^{n,m,\tilde{m}} \geqslant 0$, $\mu_3^m \geqslant 0$, $\mu_4^{n,m} \geqslant 0$, $\mu_5^{n,m} \geqslant 0$. Then, the KKT conditions are:

Stationarity:

$$\frac{\mathbf{d}f(\delta_{m,n})}{\mathbf{d}\delta_{m,n}} = P_{m,n} - \mu_1^{n,m}(1-P_{m,n}) - \sum_{\tilde{m}} \mu_2^{n,m,\tilde{m}} - \mu_3^m - \mu_4^{n,m} + \mu_5^{n,m} = 0. \tag{5.19}$$

Primal feasibility:

$$\delta_{m,n}(1-P_{m,n}) \leqslant 0, \tag{5.20}$$

$$\delta_{m,n} + \delta_{\tilde{m},n} + F_{m,\tilde{m},n} \leqslant 2, \tag{5.21}$$

$$\sum_{n=1}^{N} \delta_{m,n} \leqslant A_m, \tag{5.22}$$

$$0 \leqslant \delta_{m,n} \leqslant 1. \tag{5.23}$$

Complementary slackness:

$$\mu_1^{n,m} \delta_{m,n} (1 - P_{m,n}) = 0, \tag{5.24}$$

$$\mu_2^{n,m,\tilde{m}} \left(\delta_{m,n} + \delta_{\tilde{m},n} + F_{m,\tilde{m},n} - 2 \right) = 0, \tag{5.25}$$

$$\mu_3^m \left(\sum_{n=1}^{N} \delta_{m,n} - A_m \right) = 0, \tag{5.26}$$

$$\mu_4^{n,m} (\delta_{m,n} - 1) = 0, \tag{5.27}$$

$$\mu_5^{n,m} \delta_{m,n} = 0. \tag{5.28}$$

Let $\boldsymbol{\delta}^\diamond$ denote the optimal solution to (5.17). Then, $\boldsymbol{\delta}^\diamond$ should satisfy the KKT conditions. Suppose $\delta_{m,n}^\diamond \in (0,1)$, we have the following deduction:

$$\delta_{m,n}^\diamond \in (0,1) \begin{cases} \xrightarrow{(5.20)} P_{m,n} = 1 \\ \xrightarrow{(5.24)} \mu_1^{n,m} (1 - P_{m,n}) = 0 \\ \xrightarrow{(5.27) \text{ and } (5.28)} \mu_4^{n,m} = \mu_5^{n,m} = 0 \end{cases} \xrightarrow{(5.19)} \sum_{\tilde{m}} \mu_2^{n,m,\tilde{m}} + \mu_3^m = 1, \tag{5.29}$$

and the following properties are also obtained:

■ Suppose $\delta_{m,n}^\diamond \in (0,1)$, if for all $\tilde{m}$, $\delta_{\tilde{m},n}^\diamond \notin (0,1)$, then $\sum_{\tilde{m}} \mu_2^{n,m,\tilde{m}} = 0$ (according to (5.25)). If for all $\tilde{n} \neq n$, $\delta_{m,\tilde{n}}^\diamond \notin (0,1)$, then $\mu_3^m = 0$ (according to (5.26)). Therefore, there must exist $\tilde{m} \neq m$ or $\tilde{n} \neq n$ that $\delta_{\tilde{m},n}^\diamond \in (0,1)$ or $\delta_{m,\tilde{n}}^\diamond \in (0,1)$ (according to (5.29)).

■ Suppose $\delta_{m,n}^\diamond \in (0,1)$ and $\delta_{\tilde{m},n}^\diamond \in (0,1)$, for any $\tilde{n} \neq n$, if $\delta_{m,\tilde{n}}^\diamond \notin (0,1)$ and $\delta_{\tilde{m},\tilde{n}}^\diamond \notin (0,1)$, then $F_{m,\tilde{m},n} = 1$ (according to (5.25)).

Based on the analysis above, we find that it is difficult for $\delta_{m,n}^\diamond \in (0,1)$ to satisfy the KKT condition, which means that only a limited number of non-integer $\delta_{m,n}^\diamond$ can be obtained in the solution based on linear programming. Moreover, for the non-integer $\delta_{m,n}^\diamond$, we propose the following truncate algorithm

$$\max_{\boldsymbol{\sigma}} \sum_{m \in \Theta, n \in \Theta} \sigma_{m,n}$$

$$s.t. \begin{cases} \sum_{m \in \Theta} \sigma_{m,n} \leqslant 1 \text{ for all } n \in \Theta, \\ \sum_{n \in \Theta} \sigma_{m,n} \leqslant B_m \text{ for all } m \in \Theta, \\ \sigma_{m,n} \in \{0,1\}, \end{cases} \tag{5.30}$$

where $\Theta = \{m,n | \delta_{m,n}^\diamond \in (0,1)\}$, $\boldsymbol{\sigma} = \{\sigma_{m,n} | m \in \Theta; n \in \Theta\}$, and $B_m = \max\{\lfloor \sum_{n \in \Theta} \delta_{m,n}^\diamond \rfloor, A_m - \sum_{n \notin \Theta} \delta_{m,n}^\diamond \}$, $\lfloor Z \rfloor$ denotes the maximum integer that is no larger than Z. This is an assignment problem that can be effectively solved by many existing methods, such as the Hungarian algorithm, Auction algorithm, and so on [89].

By employing the suboptimal solution based on linear programming to solve our problem, the complexity can be largely reduced. The complexity of (5.17) is $O((MN)^3)$. For the procedure of truncating the non-integer solution to 0 or 1, we use the Hungarian algorithm to solve the assignment problem (5.30). Considering the worst situation, which means all the solutions of (5.17) are non-integer, the complexity is $O((MN)^3)$, too. Thus, the overall complexity of the problem is $O((MN)^3)$. Obviously, it can be solved in polynomial time, which means that in reality, the suboptimal solution can be obtained within several seconds. Note that in the system, the duration of each slot is at the order of a minute, so it is reasonable to assign several seconds for CBSs to calculate the suboptimal solution.

5.3 Performance Analysis and Evaluation

In this section, we will evaluate the performance of the third level spectrum utilization compared with the other two levels of spectrum utilization. Several algorithms are considered, including an optimal solution based on the branch and bound searching method, a suboptimal solution based on a single channel (sub-SC), and a suboptimal solution based on linear programming (sub-LP). Some parameters in the simulations are listed in Table 5.3. All PBSs are randomly deployed in the simulation area to use the TV channels. One channel is shared by 10 PBSs, the coverage areas of which are non-overlapped to avoid interference between each other. The deployment is shown in Figure 5.9. The PBSs with the same color share the same channel. The CBSs are deployed in the simulation area similar to the base station in the cellular network. At the beginning of the simulation, the location and the moving direction of all vehicles are generated randomly. In addition, the duration of one slot is set to be 1 minute, and the moving direction of vehicles remains unchanged during one slot and may be changed in the next slot. Fifty slots are considered, during which, if one vehicle leaves the simulation area, a new vehicle is randomly deployed in the simulation area to maintain the same number of vehicles in the same simulation scenario.

Table 5.3   Parameters for Simulations

Simulation area	$1000 \times 1000 \text{ km}^2$
Number of PBSs (L)	100
Number of primary channels (N)	10
Channel bandwidth	8 MHz
Center frequencies	$462 : 8 : 534$ MHz
TV tower transmitted power	10 kw
Noise power density	-204 dB/Hz
SNR threshold	5 dB
Threshold of the interference to CVs	-115 dBm
Threshold of the interference to PUs	-120 dBm
Vehicle average speed	350 km/h
CBS antenna height	40 m
Train antenna height	4 m

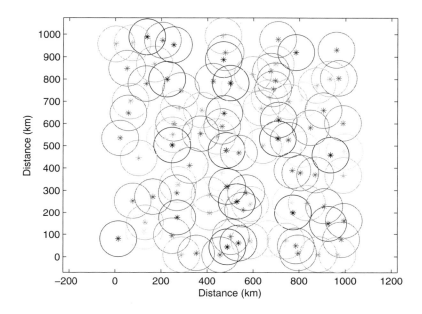

Figure 5.9: The deployment of 100 TV signal transmission towers.

First, we evaluate the complexity performance of separation computing. As discussed above, the smaller the element number of the minimum independent vehicle set (ENMIVS), the lower the overall complexity is. However, ENMIVS is dynamic and highly dependent on the distribution density of vehicles, the vehicle's location, velocity, and moving direction. Therefore, we obtain the approximate maximum ENMIVS and average ENMIVS over 100 times in Monte Carlo simulations. For comparison, we consider the baseline situation, in which separation computing is not applied. The baseline situation can be considered when ENMIVS is equal to the number of vehicles in the whole CR-HSVN. The simulation results are shown in Figure 5.10. Obviously, after performing separation computing, the ENMIVS is much smaller than the number of vehicles in the whole CR-HSVN. Thus, the computing complexity is reduced dramatically.

Figure 5.11 shows the spectrum utilization of the CR-HSVN with a different number of vehicles. We consider that the channel requirement of each vehicle is constant ($A_m = 5$, $m = 1, 2, \cdots, M$), and the minimum distance between two CBSs is 10 km. The vertical axis represents the spectrum resource of the CR-HSVN, and the horizontal axis represents the total number of vehicles in the network. As shown in Figure 5.11, both the sub-SC and sub-LP have small gaps compared with the optimal solution. The performance gap between R_1 and R_2 is caused by the activity of the

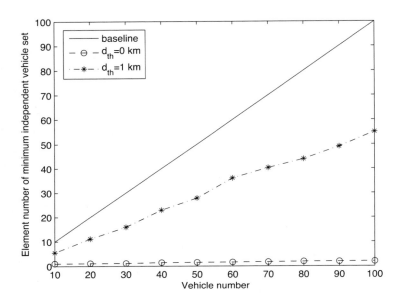

Figure 5.10: Complexity performance of separation computing.

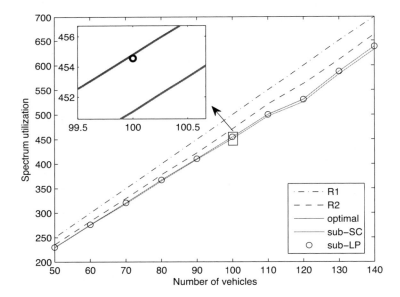

Figure 5.11: Spectrum utilization of the CR-HSVN with different vehicle numbers.

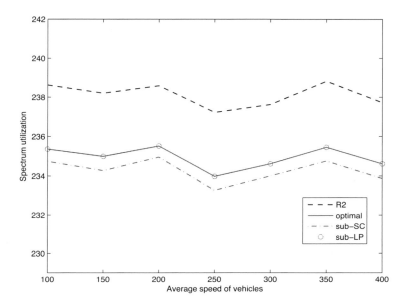

Figure 5.12: Spectrum utilization of the CR-HSVN with different average speeds of vehicles.

PUs. In addition, during certain slots, when certain vehicles cross multiple coverage of PBSs with different licensed channels, the number of licensed channels is larger than $N - A_m$, R_2 will be less than R_1 according to the definition of R_2. When the number of vehicles is small, it has a small number of vehicles that cross enough coverage of PBSs with different licensed channels. However, when the number of vehicles is large, it has a large number of vehicles that cross enough coverage of PBSs with different licensed channels. Therefore, the gap between R_1 and R_2 increases when the number of vehicles increase. The performance gap between R_2 and the optimal solution is caused by interference avoidance of the CVs. With the number of vehicles (i.e., the vehicle density) increasing, the gap between R_2 and the optimal solution increases, since high vehicle density will introduce more interference among the CVs and result in more performance losses. Figure 5.12 shows the impact of velocity, in which there are $M = 50$ vehicles with average velocity ranges from 100 to 400 km/h, and the channel requirement of each vehicle is set to be $A_m = 5$. As shown in Figure 5.12, with different average speeds of the vehicles, both the sub-SC and sub-LP have small gaps compared with the optimal solution. The proposed methods are suitable for both high-speed and low-speed scenarios.

Figure 5.13 shows the impact of d_{th}, in which there are $M = 50$ vehicles with the same channel requirement $A_m = 5$, and the minimum distance between two CBSs

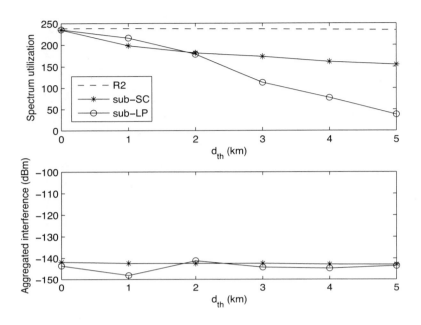

Figure 5.13: Spectrum utilization of the CR-HSVN with different d_{th}.

is 10 km. We also evaluate the maximum aggregated interference in the CR-HSVN, which is defined as

$$I = \max_{n=1,2,\cdots,N} \left\{ \max_{\mathscr{P}_l \in \mathscr{P}_{set}^n} \left\{ \sum_{m=1}^M \delta_{m,n} I_{m,\mathscr{P}_l} \right\} \right\}, \qquad (5.31)$$

where $I_{m,\mathscr{P}_l}$ denotes the interference from the vehicle $\mathscr{S}_m$ to the PUs belonging to $\mathscr{P}_l$. $I_{m,\mathscr{P}_l}$ can be approximately obtained by Equations (5.4) and (5.5). As shown in Figure 5.13, the spectrum resources utilized by the CR-HSVN are decreasing as d_{th} increases, since some vehicles cannot share the channels for the constraint of the aggregated interference when d_{th} is large. However, as d_{th} increases, the maximum aggregated interference is constrained and nearly not decreasing. This is because the distance between the vehicles is large which is in fact due to the high velocity.

Figure 5.14 depicts the spectrum utilization of the CR-HSVN with different channel requirements. In addition, Jain's fairness index [90] is employed to evaluate the fairness of our scheme and it is defined as

$$G = \frac{\left(\sum\limits_{m=1}^M \frac{x_m}{A_m} \right)^2}{M \sum\limits_{m=1}^M \left(\frac{x_m}{A_m} \right)^2}, \qquad (5.32)$$

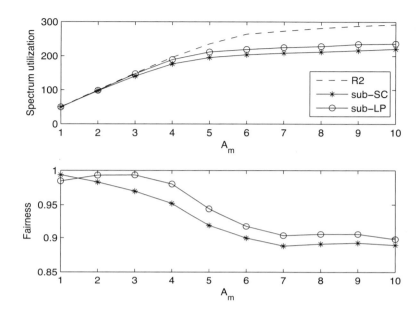

Figure 5.14: Spectrum utilization of the CR-HSVN with different channel requirements.

where x_m is the spectrum utilization allocated to the m-th vehicle. In this situation, there are $M = 50$ vehicles in the network, and the minimum distance between two CBSs is 10 km. The vertical axis represents the spectrum utilization of the CR-HSVN, and the horizontal axis represents the channel requirement of each vehicle, that is, A_m. Although, more spectrum resources can be utilized by the CR-HSVN when A_m is large, the fairness performance degrades when A_m is large.

Figure 5.15 shows spectrum utilization under different minimum distances between two CBSs in the CR-HSVN. In this situation, there are 50 vehicles in the network, and the channel requirement of each vehicle is constant ($A_m = 5$, $m = 1, 2, \cdots, M$). The vertical axis represents the spectrum utilization of the CR-HSVN, and the horizontal axis represents the minimum distance between two CBSs. This simulation shows the trade-off between performance and cost. When the minimum distance between two CBSs is large, a small number of CBSs are needed in the area. However, when the transmitted power of vehicles is large, then the interference among vehicles is increased, resulting in performance losses. Meanwhile, the maximum aggregated interference increases as the minimum distance between two CBSs increases due to the large transmitted power of vehicles.

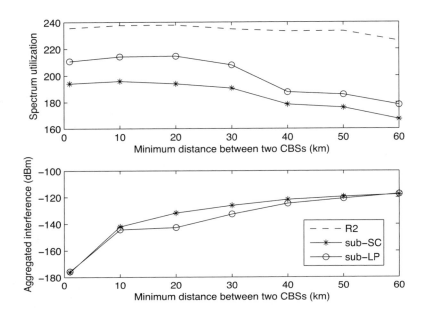

Figure 5.15: Spectrum utilization of the CR-HSVN with different minimum distances of CBS.

5.4 Summary

In this chapter, we introduced the problem of how high-speed vehicles can effectively access TV white space for broadband wireless communications. We formulated the scheduling and spectrum resource allocation problem as a linear $0-1$ integer programming optimization problem to maximize the spectrum utilization of the CR-based high-speed vehicle network. We then introduced the property of variable independence and the separation computing method to reduce the complexity dramatically without loss of optimality, and obtained the optimal solution based on the branch and bound searching method. We also introduced two suboptimal solutions with low complexity and small performance loss based on single-channel and linear programming. Simulation results showed that the considered framework offers excellent performance on spectrum utilization and fairness of the CR-based high-speed vehicle network under different situations, such as a different number of vehicles, different channel requirements, and different minimum distances of CBSs, and the aggregated interference in CR-HSVN is also very constrained.

Chapter 6

Sensing Channel Allocation in a Cognitive Radio Network for a Smart Grid

In the power grid, demand side management is considered to shape the power load for more stability, that is, during a certain period of time the electricity demand is redistributed for the balance of supply and demand. Some load shaping methods include peak clipping, valley filling, strategic conservation, and load shifting [91]. Usually, these methods can be divided into two categories. One is direct load control implemented by utilities, such as peak clipping. The other is indirect load control by motivating customers to use more electricity when the valley load appears and less electricity when the peak load appears, such as load shifting. Dynamic pricing is considered to be one of the effective motivation strategies for load shaping, especially in a smart grid, which has attracted more and more attentions due to rising energy costs and the urgent need of for reducing global carbon emissions [92]. As presented by the U.S. Department of Energy [93], a smart grid is characterized by a two-way flow of electricity and information, and will incorporate into the grid the benefits of distributed computing and communications to deliver real-time information and enable the near-instantaneous balance of supply and demand at the device level. Hence, real-time communications plays an important role in a smart grid.

As a main candidate of communication technology, wireless communications has obtained more and more attention in smart grids due to its many advantages, such as wide area coverage, cost-effectiveness, quick deployment, and so on [94–96].

However, some existing wireless communications technology suffers from several drawbacks when used in a smart grid [96]. First, the popular wireless communication standard 802.11, which employs the Industrial Scientific Medical (ISM) unlicensed spectrum bands, is not suitable for smart grids, since the ISM bands are heavily utilized in urban areas but are not suitable for the distance requirements in rural areas. Second, as the most successful commercial communication network, a cellular network is also not suitable for smart grids, since the extra expense associated with licensed bands is needed in a cellular network. Moreover, there is considerable competition for this bandwidth in urban areas and limited availability in rural areas. Third, the use of proprietary mesh network technology reduces interoperability and impedes meter diversity [96]. Hence, new wireless communication service suitable for smart grids should be investigated. Unfortunately, the spectrum scarcity is a big challenge for new wireless communications services, since the current wireless communications systems are characterized by a static spectrum allocation policy according to the spectrum allocation bodies around the world, and few spectrum resources are currently available for new wireless applications. However, a survey by Akyildiz [3] shows that current spectrum utilization is quite inefficient. Recently, dynamic spectrum access and the cognitive radio have been proposed and have obtained more and more attention for the growing need to improve spectrum utilization and solve the scarcity problem of spectrum resources [6, 7]. Some literature has tried to combine cognitive radio technology with smart grids [96–102]. Note that the cognitive radio-based network requires the enhancement of current PHY and MAC protocols to adopt spectrum-agile features, which are to allow secondary users to access the licensed spectrum band when primary users are absent. In a smart grid, all consumers communicate with utilities nearly simultaneously, which requires the high quality of the spectrum sensing scheme when cognitive radio technology is employed in a smart grid. Specifically, effective sensing channel allocation for acquiring enough available spectrum is commonly recognized as one of the most fundamental issues in a cognitive radio-based smart grid.

In this chapter, we consider the cognitive radio as the key wireless communication technology between electricity consumers and the utilities to support an effective electricity load shaping strategy in a smart grid [107]. Specifically, two sensing channel allocation strategies are introduced for the fast discovery of idle channels, which are suitable for different primary user traffic patterns.

6.1 Electricity Load Shaping Framework

6.1.1 Smart Grid Model

We will consider a time-slotted smart grid system with M distributed consumers. All consumers are independent and non-cooperative. In this chapter, we focus on the inelastic electricity demands. Generally speaking, inelastic electricity demands include lights, computers, televisions, microwave ovens, and so on. Since consumers have the habit of using such equipment, we can assume that inelastic demands at the

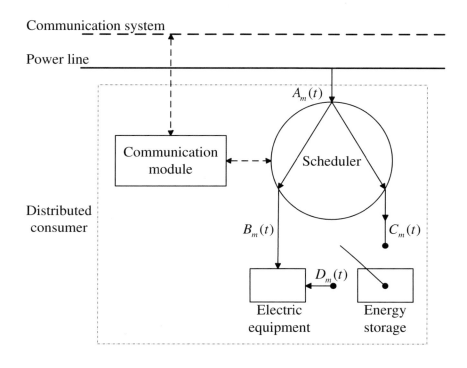

Figure 6.1: Electricity load shaping framework.

same time on different days follow the same distribution. The interaction between the utility and each consumer is shown in Figure 6.1. At the end of the $(t-1)$-th slot, $t = 2,3,\cdots$, the utility develops a pricing strategy for each consumer, which is then sent to each consumer correspondingly. After obtaining the pricing strategy, the m-th consumer, $m = 1,2,\cdots,M$, determines an action policy, including: (1) the quantity of electricity drawn from the power grid for the electric equipment during this slot, denoted as $B_m(t)$; (2) the quantity of electricity drawn from the power grid for charging the energy storage during this slot, denoted as $C_m(t)$; and (3) the quantity of electricity from its own energy storage for the electric equipment during this slot, denoted as $D_m(t)$. Let $Q_m(t)$ denote the inelastic electricity demand of the m-th consumer at slot t, then

$$Q_m(t) = B_m(t) + D_m(t), \tag{6.1}$$

where $B_m(t) \geqslant 0$ and $D_m(t) \geqslant 0$. In the following, we introduce the dynamic pricing model and discuss the energy storage model in detail.

Usually, a base load generator (BLG) is always in operation, and the cost of electricity from the BLG is relatively low. When the electricity provided by the BLG is less than the total inelastic demand, a peak load generator (PLG) should be employed, and the cost of electricity from the PLG is relatively high. However, when the electricity provided by the BLG is more than the total inelastic demand, the excess electricity is expected to be charged by the consumer's energy storage. Pricing strategy aims to encourage consumers to charge the excess electricity from the power grid when the total energy from the BLG, denoted as $G(t)$, is more than the total inelastic electricity demand, that is, $\sum_{m=1}^{M} Q_m(t)$, and to discharge the electricity for itself when $G(t)$ is less than $\sum_{m=1}^{M} Q_m(t)$. In a word, the utility expects that energy storage can replace the PLG for the energy supply, that is, the total electricity drawn from the power grid approaches $G(t)$ with as small as possible standard deviation.

We introduce a novel dynamic pricing model, called *load differentiation dynamic pricing* (LDDP). In this pricing model, the utility first obtains the load information of all consumers via the communication system. Then, let Υ denote the set of all the consumers that have successfully communicated with the utility, that is, $m \in \Upsilon$ denotes that the communication between the m-th consumer and the utility is successful, and $m \notin \Upsilon$ denotes that the communication between the m-th consumer and the utility has failed.

If $m \notin \Upsilon$, the electricity usage strategy of the m-th consumer is

$$\begin{cases} A_m(t) = B_m(t) = Q_m(t) \\ C_m(t) = D_m(t) = 0, \end{cases} \tag{6.2}$$

where $A_m(t)$ denotes the m-th consumer's total quantity of electricity drawn from the power grid. The price of the electricity with respect to the m-th consumer is set to be a constant $\bar{P}_{un}$.

If $m \in \Upsilon$, we have

$$A_m(t) = B_m(t) + C_m(t). \tag{6.3}$$

The m-th consumer is encouraged to draw a certain quantity of electricity (quota), which is denoted as $\bar{Q}_m(t)$, from the power grid. $\bar{Q}_m(t)$ can be given by

$$\bar{Q}_m(t) = G(t) \frac{H_m(t)}{\sum\limits_{m \in \Upsilon} H_m(t)}, \tag{6.4}$$

where $H_m(t)$ denotes the expected value of demand distribution of the m-th consumer at slot t. If $A_m(t) = \bar{Q}_m(t)$, the price of the electricity in the grid is set to be lowest, denoted as $\bar{P}_m(t)$. Otherwise, the price of the electricity in the grid increases as $|A_m(t) - \bar{Q}_m(t)|$ increases. We can observe that, when $\bar{Q}_m(t) < Q_m(t)$, the m-th consumer can increase $D_m(t)$ (i.e., reduce $A_m(t)$) considering (6.1). The benefit is that the consumer can use electricity in the grid at a lower electricity price and the utility can implement load shaping since less electricity is drawn from the grid. Similarly, when $\bar{Q}_m(t) \geq Q_m(t)$, the m-th consumer can increase $C_m(t)$ (i.e., increase $A_m(t)$) considering (6.3). The benefit is that the consumer can charge energy at a lower electricity price and the stored energy can be discharged when the grid electricity price is high, resulting in lower energy costs for the consumer. Meanwhile, the utility can implement load shaping since more electricity is drawn from the grid. For analysis simplicity but without loss of generality, we employ a linear function of $Am(t)$ to formulate LDDP. Then, the price of the electricity with respect to the m-th consumer is

$$f(A_m(t)) = \begin{cases} f_1(A_m(t)), & 0 \leqslant A_m(t) < \bar{Q}_m(t), \\ f_2(A_m(t)), & \bar{Q}_m(t) \leqslant A_m(t), \end{cases} \quad (6.5)$$

where $f_1(A_m(t)) = k_{1,m}(A_m(t) - \bar{Q}_m(t)) + \bar{P}_m(t)$ and $f_2(A_m(t)) = k_{2,m}(A_m(t) - \bar{Q}_m(t)) + \bar{P}_m(t)$. The parameters $k_{1,m}$ and $k_{2,m}$ indicate the encouragement strength of using energy storage for the m-th consumer. $k_{1,m} < 0$ and $k_{2,m} > 0$. Furthermore, to guarantee that the cost of using electricity, that is, $f(A_m(t)) \times A_m(t)$, is an increasing function of $A_m(t)$ when $A_m(t) > 0$, $k_{1,m}$ must satisfy $k_{1,m} \geqslant -\frac{\bar{P}_m(t)}{\bar{Q}_m(t)}$.

In addition, let V_m denote the capacity of the m-th consumer's energy storage. Due to the physical characteristics of energy storage, $C_m(t)$ and $D_m(t)$ are upper bounded by $\bar{C}_m$ and $\bar{D}_m$, respectively, that is,

$$0 \leqslant C_m(t) \leqslant \bar{C}_m, \quad (6.6)$$

$$0 \leqslant D_m(t) \leqslant \bar{D}_m. \quad (6.7)$$

Let $R_m(t)$ denote the quantity of electricity in the m-th consumer's energy storage at the beginning of the t-th slot. Then, the updated equation of $R_m(t)$ is given by

$$R_m(t) = \begin{cases} 0, & t = 1, \\ R_m(t-1) + C_m(t-1) - D_m(t-1), & t > 1. \end{cases} \quad (6.8)$$

Let $P_m(t)$ denote the average price of the electricity in the m-th consumer's energy storage at the beginning of the t-th slot. Then, the updated equation of $P_m(t)$ is given by

$$P_m(t) = \begin{cases} 0, & R_m(t) = 0, \\ \frac{\Phi(t-1) + \Psi(t-1)}{R_m(t)}, & \text{otherwise}, \end{cases} \quad (6.9)$$

where $\Phi(t-1)$ is the cost of the remaining electricity in the energy storage after discharging in the $(t-1)$-th slot and $\Psi(t-1)$ is the cost of the charged electricity in the $(t-1)$-th slot, which are given by

$$\Phi(t-1) = P_m(t-1)(R_m(t-1) - D_m(t-1)), \quad (6.10)$$

$$\Psi(t-1) = C_m(t-1)f(A_m(t-1)). \quad (6.11)$$

We also assume that charging and discharging cannot be done simultaneously, that is, $C_m(t) \neq 0 \to D_m(t) = 0$ and $D_m(t) \neq 0 \to C_m(t) = 0$, which can be formulated as

$$C_m(t)D_m(t) = 0. \quad (6.12)$$

Moreover, during each time slot, the quantity of electricity in the energy storage cannot be larger than the capacity of the storage, and cannot be smaller than zero. Therefore, $C_m(t)$ and $D_m(t)$ are also bounded by

$$0 \leqslant C_m(t) \leqslant V_m - R_m(t), \quad (6.13)$$

$$0 \leqslant D_m(t) \leqslant \min\{R_m(t), Q_m(t)\}. \quad (6.14)$$

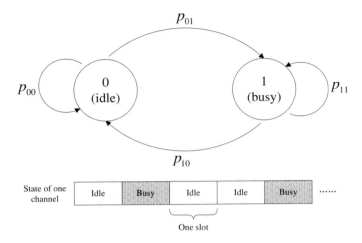

Figure 6.2: The Markov channel model.

6.1.2 The Cognitive Radio Network Model

Cognitive radio is considered to be the key wireless communication technology between consumers and utilities in this chapter. The communication architecture is discussed in the following.

First, as shown in Figure 6.2, each licensed channel used by the PU is modeled as an independent and identically distributed (i.i.d.), two-state discrete-time Markov chain [104]. The two states for each channel, "busy" (or state 1) and "idle" (or state 0), indicate the availability of transmission for consumers over that channel at a given slot. The transition probabilities are denoted as p_{ij}, $i, j = 0, 1$, where $p_{00} + p_{01} = 1$, $p_{11} + p_{10} = 1$. Note that different values of p_{ij} reflect different PU traffic patterns, which depicts the type of PU using a licensed channel. Large p_{11} and p_{00} indicate that the PU traffic is active and inactive, respectively, in a long duration. Large p_{10} and p_{01} indicate that the PU traffic is active and inactive, respectively, in a short duration. Suppose that x and y denote the amount of slots that the channel state keeps idle and busy, respectively. Then, the expectation of x and y are given by

$$E[x] = \sum_{x=1}^{+\infty} x p_{00}^{x-1} p_{01} = \frac{1}{1 - p_{00}}, \qquad (6.15)$$

$$E[y] = \sum_{y=1}^{+\infty} y p_{11}^{y-1} p_{10} = \frac{1}{1 - p_{11}}. \qquad (6.16)$$

In addition, the probability of a channel being idle, P_a, which is defined as the fraction of time in which a channel is in an idle state, is given by [105]

$$P_a = \frac{E[x]}{E[x] + E[y]}. \qquad (6.17)$$

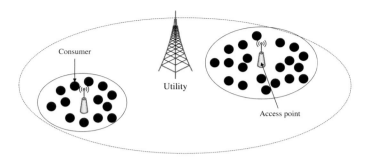

Figure 6.3: The considered communication architecture.

In a smart grid, effective communications between consumers and the utility must be guarded. Limited by the transmission range and the contradiction between a large number of consumers and limited spectrum resources, it cannot be realized that all consumers simultaneously and directly communicate with the utility. Hence, similar to Fatemieh, Chandra, and Gunter [96], the clustered communication architecture is employed. As shown in Figure 6.3, all consumers are divided into many clusters, each of which has a cluster head, called an access point. Slot-based communication is employed, and the slot structure is shown in Figure 6.4. In each slot, each access point first senses the licensed band during the sub-slot s_1 to discover the idle channels, then collects the information of the consumers in its own communication range during the sub-slot s_2 and sends the collected information to the utility during the sub-slot s_3 over the discovered idle channels for the load shaping strategy in the next slot. Since the amount of each consumer's information is small, it is reasonable to assume that the duration of s_1 is far larger than the duration of s_2 and s_3.

During the sub-slot s_1, energy detection, which is a simple and effective method for detection of unknown signals, is employed for spectrum sensing. Based on the discussion in Liang et al. [20], the sensing time for detecting one licensed channel in

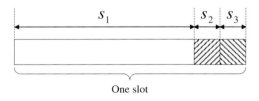

Figure 6.4: The slot structure.

energy detection can be given by

$$\tau = \frac{\left(\mathscr{Q}^{-1}(P_f) - \mathscr{Q}^{-1}(1 - P_m)\sqrt{2\gamma + 1}\right)^2}{\gamma^2 f_s}, \tag{6.18}$$

where γ is the received signal-to-noise ratio (SNR), f_s is the sampling frequency, $P_f = \Pr(D_1|H_0)$ denotes the probability of missed detection, $P_m = \Pr(D_0|H_1)$ denotes the probability of a false alarm, H_0 and H_1 represent that the licensed channel is idle and busy, respectively, D_0 and D_1 represent that the licensed channel is detected as being idle and busy, respectively, $\mathscr{Q}^{-1}(\cdot)$ is the inverse function of $\mathscr{Q}(\cdot)$, and $\mathscr{Q}(\cdot)$ is the complementary distribution function of the standard Gaussian distribution, that is,

$$\mathscr{Q}(x) = \frac{1}{\sqrt{2\pi}} \int_x^\infty \exp\left(-\frac{t^2}{2}\right) dt. \tag{6.19}$$

In addition, the posterior probability $\Pr(H_0|D_0)$ reflects the reliability of communication. When $\Pr(H_0|D_0) = 1$, we consider that the communication is completely reliable. Note that the communication errors resulting from noise, fading, shadowing, and so on, are omitted in this chapter, since they are considered to be minimized via mature communication technology. According to Bayes' theorem, $\Pr(H_0|D_0)$ can be obtained by

$$\begin{aligned} \Pr(H_0|D_0) &= \frac{\Pr(D_0|H_0)\Pr(H_0)}{\Pr(D_0|H_0)\Pr(H_0) + \Pr(D_0|H_1)\Pr(H_1)} \\ &= \frac{1}{1 + \frac{P_m(1-P_a)}{(1-P_f)P_a}}. \end{aligned} \tag{6.20}$$

6.2 Sensing Channel Allocation and Load Shaping Strategies

6.2.1 Sensing Channel Allocation Strategies

Two sensing channels allocation strategies are introduced to the fast discovery of idle channels during the sensing time in each slot, as shown in Figure 6.5. Let T_s denote the total sensing time in each slot, $T_s \leqslant s_1$. In the first strategy, called the *Random* allocation strategy, the access point randomly selects licensed channels to sense in each slot. Thus, during T_s, the expectation of the total number of discovered channels can be given by

$$N_1 = \frac{T_s}{\tau} P_a. \tag{6.21}$$

In the second strategy, called the *Keep-on* allocation strategy, the access point first senses the idle channels detected in the former slot. Then, the access point randomly selects other channels to sense. Thus, during T_s, the expectation of the total number

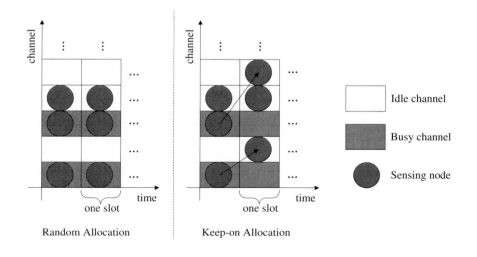

Figure 6.5: Sensing channel allocation strategies.

of detected idle channels can be given by

$$T_s = N_2\tau + N_2 p_{01}\frac{\tau}{P_a}$$

$$\Rightarrow N_2 = \frac{T_s}{\tau}\frac{P_a}{P_a + p_{01}}. \tag{6.22}$$

6.2.2 Load Shaping Strategy

Next, we consider effective load shaping strategy. First, we introduce the concept of the *experience function* of a consumer, denoted as $U(x, w)$, where x is the electricity consumption level of the consumer, and w is a parameter which may vary among consumers and also at different time durations. $U(x, w)$ represents the level of satisfaction obtained by the consumer as a function of its electricity consumption. In addition, when $Q_m(t) < R_m(t)$, the consumer may not draw electricity from the power grid, and only uses the electricity from its own energy storage. This case is not expected by the utility since the total valley load may be reduced further in this case. Thus, we introduce an auxiliary variable, which can be used by the utility to impact the operation of energy storage. Specifically, when the power demand is lower than the given quota for a consumer, increasing the parameter $\mu(t)$ in the auxiliary variable can encourage the consumer to draw more electricity from the power grid at a lower electricity price by charging energy, and the stored energy can be discharged when the grid electricity price is high, resulting in a lower energy cost for the consumer. Meanwhile, the utility can implement load shaping since more electricity is drawn from the grid. In a word, auxiliary variables can help to achieve a win-win situation for both the utility and consumers.

We define the welfare function of the m-th consumer as

$$W_m(t) = \underbrace{U(Q_m(t), w_m)}_{\text{profit if using electricity}}$$

$$+ \underbrace{R_m(t+1)P_m(t+1) - R_m(t)P_m(t)}_{\text{profit due to the change of energy in storage}}$$

$$- \underbrace{B_m(t)f(A_m(t))}_{\text{cost if using electricity}} \underbrace{- C_m(t)f(A_m(t))}_{\text{cost if charging}}$$

$$- \underbrace{\mu(t)D_m(t)P_m(t)}_{\text{auxiliary variable}}, \qquad (6.23)$$

where $\mu(t) \geq 0$ and its value can be dynamically designed online, which will be investigated later.

Substituting (6.8) and (6.9) into (6.23), the welfare function can be rewritten as

$$W_m(t) = U(Q_m(t), w_m) - B_m(t)f(A_m(t))$$
$$- (1 + \mu(t))D_m(t)P_m(t). \qquad (6.24)$$

Each consumer aims to maximize its own welfare. Note that one of the advantages of the proposed load shaping strategy is to guarantee the inelastic electricity demand of the consumers, that is, $U(Q_m(t), w_m)$ is independent of the consumer's action policy. Therefore, maximizing the welfare of the m-th consumer can be written as

$$\min_{B_m(t), C_m(t), D_m(t)} \lim_{T \to +\infty} \frac{1}{T} \sum_{t=1}^{T} \Theta(B_m(t), C_m(t), D_m(t)) \qquad (6.25)$$

$$s.t. \ (6.1), \ (6.6), \ (6.7), \ (6.8),$$
$$(6.9), \ (6.12), \ (6.13), \ (6.14).$$

where $\Theta(B_m(t), C_m(t), D_m(t)) = B_m(t)f(B_m(t) + C_m(t)) + (1 + \mu(t))D_m(t)P_m(t)$.

Obviously, the action policy in the $(t+1)$-th slot is related to the action policy in slot t and the price information in the $(t+1)$-th slot. However, when the action policy in slot t should be determined, the price information in the $(t+1)$-th slot is unknown, thus, it is not clear how the action policy in slot t will affect the action policy in the $(t+1)$-th slot. Therefore, it is very difficult to obtain the optimal solution to (6.25). We then implement an effective greedy strategy to obtain solutions to (6.26)–(6.29) in a slot-by-slot fashion, which are suboptimal but practical solutions to (6.25). The per-slot optimization problem can be formulated as

$$\min_{B_m(t), C_m(t)} \Omega(B_m(t), C_m(t)) \qquad (6.26)$$

$$s.t. \ Q_m(t) - \beta \leqslant B_m(t) \leqslant Q_m(t), \qquad (6.27)$$
$$0 \leqslant C_m(t) \leqslant \alpha, \qquad (6.28)$$
$$C_m(t)(Q_m(t) - B_m(t)) = 0, \qquad (6.29)$$

where $\Omega(B_m(t), C_m(t)) = B_m(t)f(B_m(t)+C_m(t)) + (1+\mu(t))P_m(t)(Q_m(t)-B_m(t))$, $\alpha = \min\{\bar{C}_m, V_m - R_m(t)\}$, $\beta = \min\{\bar{D}_m, R_m(t), Q_m(t)\}$, and $D_m(t)$ are replaced by $Q_m(t) - B_m(t)$. The optimal closed-form solution to (6.26)–(6.29) can be easily obtained by using an analytic function. After the action policy in slot t is obtained, the state of the energy storage is updated by (6.8) and (6.9).

In addition, how to design $\mu(t)$ is a very important issue in the proposed strategy. As mentioned above, the utility expects that the total electricity drawn from the power grid approaches $G(t)$ with a small standard deviation, which can be formulated as

$$\min_{\mu(t)} \left(G(t) - \sum_{m=1}^{M} (B_m^*(t) + C_m^*(t)) \right)^2 \tag{6.30}$$

$$s.t. \ \mu(t) \geqslant 0,$$

where $B_m^*(t)$ and $C_m^*(t)$ are the optimal action policy of the m-th consumer. To obtain the optimal $\mu(t)$, we introduce an important property first.

Property 6.1 $A_m^*(t) = B_m^*(t) + C_m^*(t)$ *is a non-decreasing function with respect to* $\mu(t)$, *that is,* $\forall \ \mu_\diamond(t) > \mu(t)$, $A_{m,\diamond}^*(t) \geqslant A_m^*(t)$, *where* $A_{m,\diamond}^*(t) = B_{m,\diamond}^*(t) + C_{m,\diamond}^*(t)$ *and* $A_m^*(t) = B_m^*(t) + C_m^*(t)$ *are the optimal action policies of the m-th consumer under* $\mu_\diamond(t)$ *and* $\mu(t)$, *respectively.*

Let $Y(t) = \sum_{m=1}^{M}(B_m^*(t) + C_m^*(t))$. Based on Property 6.1, $Y(t)$ is also a non-decreasing function with respect to $\mu(t)$. Moreover, the object function of (6.30) is a parabolic function with respect to $Y(t)$. Therefore, the optimal $\mu(t)$ is unique, and the approximate optimal $\mu(t)$ in each slot can be obtained by a numerical algorithm (see Algorithm 6.1).

Note that *step* is the step length when searching $\mu(t)$. In addition, as discussed above, $Y(t)$ may be kept unchanged as $\mu(t)$ is increasing. In this situation, K^{max} is introduced to avoid an infinite loop. In fact, when the *step* is small enough and K^{max} are large enough, we can obtain the optimal $\mu(t)$. However, a small *step* and large K^{max} result in high-computing complexity.

6.3 Performance Analysis and Evaluation

6.3.1 Performance of Sensing Channel Allocation

To evaluate the communication system in a CR-based smart grid, we introduce the performance of communication reliability and fast discovery of idle channels in the following.

Figure 6.6 shows the performance of communication reliability with different PU traffic patterns. When $p_{00} = 0.1$ and $p_{11} = 0.9$, that is, $P_a = 0.1$, very small P_f and P_m can guarantee a large value of $\Pr(H_0|D_0)$, which increases the cost of spectrum sensing. On the contrary, when $p_{00} = 0.9$ and $p_{11} = 0.1$, that is, $P_a = 0.9$, a large value of $\Pr(H_0|D_0)$ can be guaranteed even though P_f and P_m is very large. Note that large values of P_f and P_m result in the simple design of a spectrum sensing algorithm.

Algorithm 6.1 The pseudo-code for obtaining the approximate optimal $\mu(t)$

Procedure:
1: Initialize $\mu(t) = 0$, $X_0 = \infty$, $X_1 = \infty$, $K_e = 0$, $K_c = 1$;
2: while $K_c = 1$
3:     Obtain $B_m^*(t)$ and $C_m^*(t)$ according to (6.26), $m = 1, 2, \cdots, M$
4:     $X_0 = X_1$;
5:     $X_1 = Y(t)$;
6:     if $X_1 \leqslant X_0$ and $K_e < K^{\max}$
7:         $\mu(t) = \mu(t) + step$;
8:         if $X_1 = X_0$
9:             $K_e = K_e + 1$;
10:        else
11:            $K_e = 0$;
12:        end if
13:    else
14:        $\mu(t) = \mu(t) - step$;
15:        $K_c = 0$;
16:    end if
17: end while

Figures 6.7 and 6.8 show the performance of fast discovery of idle channels, where $P_m = P_f$. As shown in Figure 6.7, the number of detected idle channels increases when P_f is increasing. When $p_{00} = p_{11} = 0.9$, that is, the state of the licensed channel changes slowly, the *Keep-on* allocation strategy outperforms the *Random*

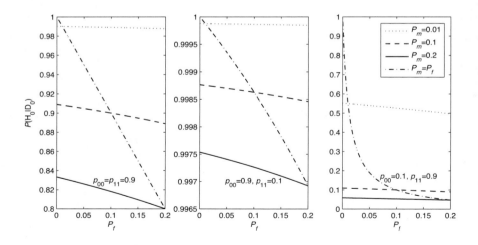

Figure 6.6: The performance of communication reliability with different PU traffic patterns.

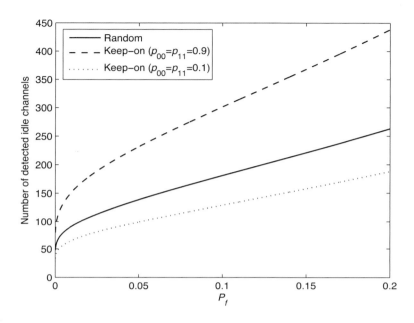

Figure 6.7: The performance of fast discovery of idle channels with different P_f.

allocation strategy. However, when $p_{00} = p_{11} = 0.1$, the *Random* allocation strategy outperforms the *Keep-on* allocation strategy since the state of the licensed channel changes rapidly in this situation. Figure 6.8 illustrates the performance of the two sensing channel allocation strateges with different PU traffic patterns, that is, different values of p_{00} and p_{11}.

6.3.2 Performance of Electricity Load Shaping

We then evaluate the performance of our load shaping strategy with Monte Carlo searching. For all simulations, we consider that $\bar{P}_{un} = 1.0$ RMB/kWh, $\bar{P}(t) = 0.5$ RMB/kWh, $G(t) = 10$ kWh. The duration of one slot is set to be 10 minutes. There are $M = 100$ consumers, which is divided into two classes. The statistical properties of $Q_m(t)$ remain unchanged in different slots. If the m-th consumer belongs to class 1, $Q_m(t)$ follows the truncated normal distribution with mean 0.05, variance 0.02, and range $(0, 0.2)$, $\bar{C}_m = \bar{D}_m = 0.04$ kWh, $V_m = V_{m,1}$. If the m-th consumer belongs to class 2, $Q_m(t)$ follows the truncated normal distribution with mean 0.15, variance 0.07, and range $(0, 0.6)$, $\bar{C}_m = \bar{D}_m = 0.12$ kWh, $V_m = 3V_{m,1}$. Note that the effectiveness of our strategy is independent of the distribution type of $Q_m(t)$.

Figure 6.9 shows the performance of load shaping over time. Obviously, the load is reshaped, the fluctuation of the load is mitigated dramatically, and the mean of the

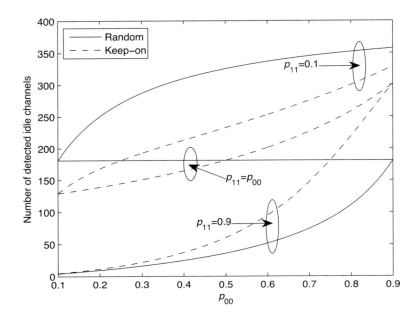

Figure 6.8: The performance of fast discovery of idle channels with different PU traffic patterns.

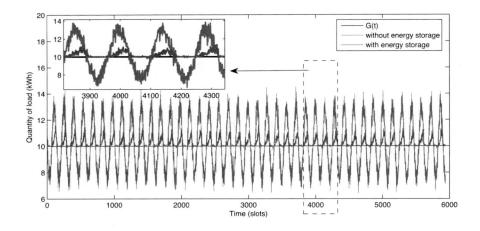

Figure 6.9: Total electricity load over time slots.

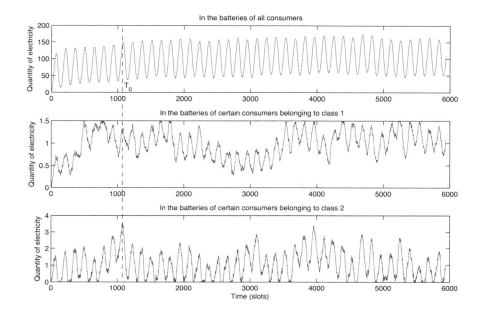

Figure 6.10: Quantity of electricity in the batteries over time slots.

load approaches the total energy from the BLG, that is, $G(t)$. Figure 6.10 shows the quantity of electricity in the batteries over time slots. Note that the system remains stable after a certain time slot, denoted as T_0, since $R_m(1) = 0$ results in charging with a large probability at the former time slots. Therefore, we will evaluate the performance of the novel load shaping strategy starting with T_0 in the following.

The consumer's cost results from purchasing electricity from the power grid. When we do not consider the energy storage, all the inelastic electricity demand must be provided by the power grid, the cost of the m-th consumer, averaged over time, is given by

$$\Gamma_{\text{wo},m} = \lim_{T \to +\infty} \frac{1}{T - T_0} \sum_{t=T_0}^{T} Q_m(t) f(Q_m(t)). \tag{6.31}$$

When we consider the energy storage, the purchasing electricity from the power grid is $A_m(t)$ for the m-th consumer, and the cost, averaged over time, is given by

$$\Gamma_{\text{w},m} = \lim_{T \to +\infty} \frac{1}{T - T_0} \sum_{t=T_0}^{T} A_m(t) f(A_m(t)). \tag{6.32}$$

Then, the percentage of savings with respect to the m-th consumer's cost is obtained by

$$\rho_m = \frac{\Gamma_{\text{wo},m} - \Gamma_{\text{w},m}}{\Gamma_{\text{wo},m}} \times 100\%, \tag{6.33}$$

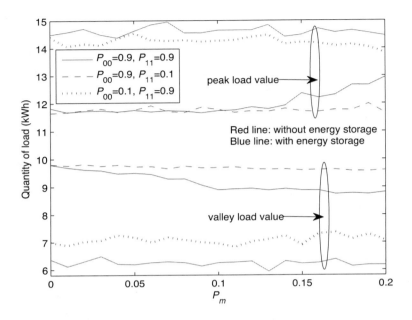

Figure 6.11: Peak/valley load performance versus P_m and P_f.

and the percentage of savings with respect to all of the consumer's cost is obtained by

$$\rho = \frac{\sum\limits_{m=1}^{M} \Gamma_{\text{wo},m} - \sum\limits_{m=1}^{M} \Gamma_{\text{w},m}}{\sum\limits_{m=1}^{M} \Gamma_{\text{w},m}} \times 100\%. \tag{6.34}$$

The performance of load shaping versus P_m and P_f is illustrated in Figures 6.11, 6.12, and 6.13. In these simulations, $P_f = P_m$. Obviously, small values of P_m and P_f result in good performance of the load shaping strategy. In addition, the PU traffic pattern also influences the performance of the load shaping strategy. When $p_{00} = 0.9$ and $p_{11} = 0.1$, there are enough idle channels for consumer communication, resulting in good performance of the load shaping strategy. When $p_{00} = 0.1$ and $p_{11} = 0.9$, there are not enough idle channels for consumer communication. Therefore, the performance of the load shaping strategy decreases dramatically. In the above two situations, P_m and P_f have little impact on the performance of the load shaping strategy. However, when $p_{00} = p_{11} = 0.9$, that is, the probability of a channel being idle is $P_a = 0.5$, the performance of the load shaping strategy is sensitive to the value of P_m and P_f.

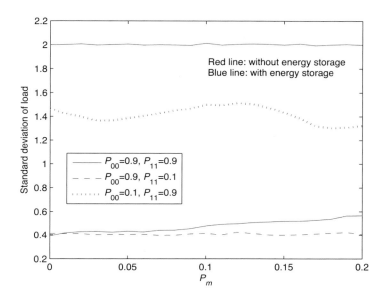

Figure 6.12: Standard deviation of the load versus P_m and P_f.

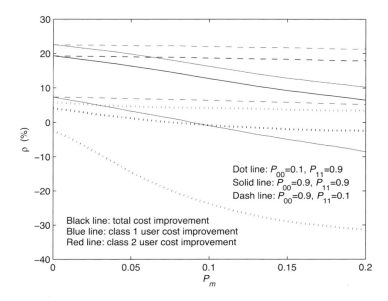

Figure 6.13: The percentage of savings with respect to the consumer's cost versus P_m and P_f.

6.4 Summary

In this chapter, we introduced the application of sensing channel allocation in a cognitive radio network for a smart grid. In a cognitive radio network for a smart grid, the communication scenario has some new characteristics. We introduced two sensing channel allocation strategies to obtain enough available channels for the communication between the utility and electricity consumer. Simulation results show that primary user traffic patterns influence the performance of communication reliability and fast available channel discovery. The accuracy of spectrum sensing also influences the performance of fast available channel discovery. The effects of spectrum sensing errors and the PU traffic patterns on the performance of the novel load shaping strategy are also evaluated.

References

1. United States frequency allocation chart. Accessed on September 10, 2012 as www.ntia.doc.gov/page/2011/united-states-frequency-allocation-chart.

2. T. He, "Establish the sense of crisis of spectrum scarcity under the pressure of future dataflow" [Online]. Accessed on March 16, 2012 as www.cww.net.cn/cwwMag/html/2011/2/21/201122194281407.htm.

3. I. F. Akyildiz, W. Y. Lee, M. C. Vuran, and S. Mohanty, "NeXt generation/dynamic spectrum access/cognitive radio wireless networks: A survey," *Computer Networks*, vol. 50, no. 13, pp. 2127–2159, September 2006.

4. M. Islam, C. L. Koh, S. W. Oh, X. Qing, Y. Y. Lai, C. Wang, Y-C. Liang, B. E. Toh, F. Chin, G. L. Tan, and W. Toh, "Spectrum survey in Singapore: Occupancy measurements and analyses," *Proc. International Conference on Cognitive Radio Oriented Wireless Networks and Communications*, pp. 1–7, May 2008.

5. Federal Communications Commission, "Spectrum policy task force," Report ET Docket No. 02-135, November 2002.

6. S. Haykin, "Cognitive radio: Brain-empowered wireless communications," *IEEE Journal on Selected Areas in Communications*, vol. 23, no. 2, pp. 201–220, February 2005.

7. J. Mitola, "Cognitive radio for flexible mobile multimedia communications," *Proc. IEEE Mobile Multimedia Conference*, pp. 3–10, November 1999.

8. J. Mitola et al., "Cognitive radio: Making software radios more personal," *IEEE Personal Communications*, vol. 6, no. 4, pp. 13–18, August 1999.

9. J. Mitola, "Cognitive radio: An integrated agent architecture for software defined radio," Doctor of Technology, Royal Inst. Technol. (KTH), Stockholm, Sweden, 2000.

10. IEEE 802.22, Working Group on Wireless Regional Area Networks (WRAN) [Online]. Accessed on March 14, 2011 as www.ieee802.org/22/.

11. T. Luo, T. Jiang, W. Xiang, and H.-H. Chen, "A subcarriers allocation scheme for cognitive radio systems based on multi-carrier modulation," *IEEE Transactions on Wireless Communications*, vol. 7, no. 9, pp. 3335–3340, September 2008.

12. Q. Zhao and B. M. Sadler, "A survey of dynamic spectrum access," *IEEE Signal Processing Magazine*, vol. 24, no. 3, pp. 79–89, May 2007.

13. Q. Zhao, S. Geirhofer, L. Tong, and B. M. Sadler, "Optimal dynamic spectrum access via periodic channel sensing," *Proc. IEEE Wireless Communications and Networking Conference*, pp. 33–37, March 2007.

14. Q. Zhao, L. Tong, A. Swami, and Y. Chen, "Decentralized cognitive MAC for opportunistic spectrum access in ad hoc networks: A POMDP framework," *IEEE Journal on Selected Areas in Communications*, vol. 25, no. 3, pp. 589–600, April 2007.

15. Y. Chen, Q. Zhao, and A. Swami, "Joint design and separation principle for opportunistic spectrum access in the presence of sensing errors," *IEEE Transactions on Information Theory*, vol. 54, no. 5, pp. 2053–2071, May 2008.

16. N. B. Chang and M. Liu, "Competitive analysis of opportunistic spectrum access strategies," *Proc. IEEE INFOCOM*, pp. 1–9, April 2008.

17. D. Qu, Z. Wang, and T. Jiang, "Extended active interference cancellation for sidelobe suppression in cognitive radio OFDM systems with cyclic prefix," *IEEE Transactions on Vehicular Technology*, vol. 59, no. 4, pp. 1689–1695, May 2010.

18. R. Zhang and Y.-C. Liang, "Exploiting multi-antennas for opportunistic spectrum sharing in cognitive radio networks," *IEEE Journal on Selected Topics in Signal Processing*, vol. 2, no. 1, pp. 88–102, February 2008.

19. S. Huang, X. Liu, and Z. Ding, "Optimal sensing-transmission structure for dynamic spectrum access," *Proc. IEEE INFOCOM*, pp. 1–9, April 2009.

20. Y.-C. Liang, Y. Zeng, E. Peh, and A. T. Hoang, "Sensing-throughput tradeoff for cognitive radio networks," *IEEE Transactions on Wireless Communications*, vol. 7, no. 4, pp. 1326–1337, April 2008.

21. A. T. Hoang, Y.-C. Liang, D. T. C. Wong, Y. Zeng, and R. Zhang, "Opportunistic spectrum access for energy-constrained cognitive radios," *IEEE Transactions on Wireless Communications*, vol. 8, no. 3, pp. 1206–1211, March 2009.

22. S. Zheng, Y.-C. Liang, P. Kam, and A. T. Hoang, "Cross-layered design of spectrum sensing and MAC for opportunistic spectrum access," *Proc. IEEE Wireless Communications and Networking Conference*, pp. 1–6, March 2009.

23. Y. Cao, D. Qu, and T. Jiang, "Throughput maximization in cognitive radio system with transmission probability scheduling and traffic pattern prediction," *Mobile Networks and Applications*, vol. 17, no. 5, pp. 604–617, October 2012.

24. S. Huang, X. Liu, and Z. Ding, "Optimal transmission strategies for dynamic spectrum access in cognitive radio networks," *IEEE Transactions on Mobile Computing*, vol. 8, no. 12, pp. 1636–1648, December 2009.

25. Y. Fang, "Hyper-Erlang distribution model and its application in wireless mobile networks," *Wireless Networks*, vol. 7, no. 3, pp. 211–219, May 2001.

26. F. Barcelo and J. I. Sanchez, "Probability distribution of the inter-arrival time to cellular telephony channels," *Proc. IEEE Vehicular Technology Conference*, pp. 762–766, May 1999.

27. S. Geirhofer, L. Tong, and B. M. Sadler, "Dynamic spectrum access in the time domain: Modeling and exploiting white space," *IEEE Communications Magazine*, vol. 45, no. 5, pp. 66–72, May 2007.

28. A. Schrijver, *Theory of Linear and Integer Programming*. New York: John Wiley & Sons, 1998.

29. I. A. Akbar and W. H. Tranter, "Dynamic spectrum allocation in cognitive radio using hidden Markov models: Poisson distributed case," *IEEE Southeast Conference*, pp. 196–201, 2007.

30. L. R. Rabiner, "A tutorial on hidden Markov models and selected application in speech recognition," *Proceedings of the IEEE*, vol. 77, no. 2, pp. 257–286, February 1989.

31. C. Ghosh, C. Cordeiro, D. P. Agrawal, and M. B. Rao, "Markov chain existence and hidden Markov models in spectrum sensing," *Proc. IEEE International Conference on Pervasive Computing and Communications*, pp. 1–6, March 2009.

32. W. Turin, *Digital Transmission Systems: Performance Analysis and Modeling*. New York: McGraw-Hill, 1998.

33. W. Nam, W. Chang, S.-Y Chung, and Y. Lee, "Transmit optimization for relay-based cellular OFDMA systems," *Proc. IEEE International Conference on Communications*, pp. 5714–5719, June 2007.

34. M. Kim and H. Lee, "Radio resource management for a two-hop OFDMA relay system in downlink," *Proc. IEEE Symposium on Computers and Communications*, pp. 25–31, July 2007.

35. S. Kim, X. Wang, and M. Madihian, "Optimal resource allocation in multi-hop OFDMA wireless networks with cooperative relay," *IEEE Transactions on Wireless Communications*, vol. 7, no. 5, pp. 1833–1838, May 2008.

36. K. Sundaresan and S. Rangarajan, "Efficient algorithms for leveraging spatial reuse in OFDMA relay networks," *Proc. IEEE INFOCOM*, pp. 1539–1547, April 2009.

37. H. Xu and B. Li, "XOR-assisted cooperative diversity in OFDMA wireless networks: Optimization framework and approximation algorithms," *Proc. IEEE INFOCOM*, pp. 2141–2149, April 2009.

38. X. Li, T. Jiang, S. Cui, J. An, and Q. Zhang, "Cooperative communications based on rateless network coding in distributed MIMO systems," *IEEE Wireless Communications Magazine*, vol. 17, no. 3, pp. 60–67, June 2010.

39. S. Kim, W. Choi, Y. Choi, J. Lee, Y. Han, and I. Lee, "Downlink performance analysis of cognitive radio-based cellular relay networks," *Proc. International Conference on Cognitive Radio Oriented Wireless Networks and Communications*, pp. 1–6, May 2008.

40. R. Wang, V. Lau, and Y. Cui, "Decentralized fair scheduling in two-hop relay-assisted cognitive OFDMA systems," *IEEE Journal of Selected Topics in Signal Processing*, vol. 5, no. 1, pp. 171–181, January 2011.

41. J. Sachs, I. Mari, and A. Goldsmith, "Cognitive cellular networks within the TV spectrum," *Proc. IEEE International Symposium on New Frontiers in Dynamic Spectrum Access Networks*, pp. 1–8, April 2010.

42. T. Luo, F. Lin, T. Jiang, M. Guizani, and W. Chen, "Multicarrier modulation and cooperative communication in multi-hop cognitive radio networks," *IEEE Wireless Communications Magazine*, vol. 18, no. 1, pp. 38–45, February 2011.

43. H. V. Poor, *An Introduction to Signal Detection and Estimation*. New York: Springer-Verlag, 1994.

44. S. M. Kay, *Fundamentals of Statistical Signal Processing: Detection Theory*. Englewood Cliffs, NJ: Prentice Hall, 1998.

45. S. Enserink and D. Cochran, "A cyclostationary feature detector," *Proc. Asilomar Conference on Signals, Systems, and Computers*, pp. 806–810, November 1994.

46. Y. Xin, H. Zhang, and S. Rangarajan, "SSCT: A simple sequential spectrum sensing scheme for cognitive radio," *Proc. IEEE Global Telecommunications Conference*, pp. 1–6, November 2009.

47. Y. Shei and Y. T. Su, "A sequential test-based cooperative spectrum sensing scheme for cognitive radios," *Proc. IEEE International Symposium on Personal, Indoor and Mobile Radio Communications*, pp. 1–5, September 2008.

48. A. T. Hoang, Y. C. Liang, and Y. Zeng, "Adaptive joint scheduling of spectrum sensing and data transmission in cognitive radio networks," *IEEE Transactions on Communications*, vol. 58, no. 1, pp. 235–246, January 2010.

49. H. Kim and K. G. Shin, "Fast discovery of spectrum opportunities in cognitive radio networks," *Proc. IEEE International Symposium on New Frontiers in Dynamic Spectrum Access Networks*, pp. 1–12, October 2008.

50. S. J. Kim and G. B. Giannakis, "Sequential and cooperative sensing for multichannel cognitive radios," *IEEE Transactions on Signal Processing,* vol. 58, no. 8, pp. 4239–4253, August 2010.

51. J. Oksanen, J. Lunden, and V. Koivunen, "Reinforcement learning based sensing policy optimization for energy efficient cognitive radio networks," *Neurocomputing,* vol. 80, pp. 102–110, March 2012.

52. Z. Wang, D. Qu, and T. Jiang, "Novel adaptive collaboration sensing for efficient acquisition of spectrum opportunities in cognitive radio networks," *Wireless Networks*, DOI: 10.1007/s11276-012-0463-y.

53. V. Fodor, I. Glaropoulos, and L. Pescosolido, "Detecting low-power primary signals via distributed sensing to support opportunistic spectrum access," *Proc. IEEE International Conference on Communications,* pp. 1–6, June 2009.

54. F. F. Digham, M. S. Alouini, and M. K. Simon, "On the energy detection of unknown signals over fading channels," *IEEE Transactions on Wireless Communications,* vol. 55, no. 1, pp. 21–24, January 2007.

55. Z. Quan, S. Cui, H. Poor, and A. Sayed, "Collaborative wideband sensing for cognitive radios," *IEEE Signal Processing Magazine,* vol. 25, no. 6, pp. 60–73, November 2008.

56. A. Wald, "Sequential tests of statistical hypothesis," *The Annals Math. Statistics*, vol. 16, no. 2, pp. 117–186, June 1945.

57. R. Bellman, *Applied Dynamic Programming*. Princeton, NJ: Princeton University Press, 1962.

58. R. E. Larson and J. L. Casti, *Principles of Dynamic Programming. Part I: Basic Analytic and Computational Methods*. New York: Marcel Dekker, 1978.

59. Truncated normal distribution [Online]. Accessed on February 23, 2012 as http://en.wikipedia.org/wiki/Truncated_normal_distribution.

60. E. Peh and Y. C. Liang, "Optimization for cooperative sensing in cognitive radio networks," *Proc. IEEE Wireless Communication and Networking Conference,* pp. 27–32, March 2007.

61. J. Laneman, D. Tse, and G. Wornell, "Cooperative diversity in wireless networks: Efficient protocols and outage behavior," *IEEE Transactions on Information Theory*, vol. 50, no. 12, pp. 3062–3080, December 2004.

62. Y. Yang, H. Hu, J. Xu, and G. Mao, "Relay technologies for WiMAX and LTE-advanced mobile systems," *IEEE Communications Magazine*, vol. 47, no. 10, pp. 100–105, October 2009.

63. S. Kadloor and R. Adve, "Relay selection and power allocation in cooperative cellular networks," *IEEE Transactions on Wireless Communications*, vol. 9, no. 5, pp. 1676–1685, May 2010.

64. Y. Cao, T. Jiang, C. Wang, and L. Zhang, "CRAC: Cognitive radio assisted cooperation for downlink transmissions in OFDMA-based cellular networks," *IEEE Journal on Selected Areas in Communications*, vol. 30, no. 9, pp. 1614–1622, October 2012.

65. J. Shen, T. Jiang, S. Liu, and Z. Zhang, "Maximum channel throughput via cooperative spectrum sensing in cognitive radio networks," *IEEE Transactions on Wireless Communications*, vol. 8, no. 10, pp. 5166–5175, October 2009.

66. J. Proakis, *Digital Communications*, 3rd edition. New York: McGraw-Hill, 1995.

67. A. Goldsmith and S. Chua, "Variable-rate variable-power MQAM for fading channels," *IEEE Transactions on Communications*, vol. 45, no. 10, pp. 1218–1230, October 1997.

68. "Multi-hop relay system evaluation methodology" [Online]. Accessed at: http://ieee802.org/16/relay/docs/80216j-06_013r3.pdf

69. T. Ng and W. Yu, "Joint optimization of relay strategies and resource allocations in cooperative cellular networks," *IEEE Journal on Selected Areas in Communications*, vol. 25, no. 2, pp. 328–339, February 2007.

70. W. Yu and R. Lui, "Dual methods for nonconvex spectrum optimization of multicarrier systems," *IEEE Transactions on Communications*, vol. 54, no. 7, pp. 1310–1322, July 2006.

71. S. Boyd, L. Xiao, and A. Mutapcic, "Subgradient methods," Lecture Notes of EE364b, Stanford University, Spring Quarter 2007–2008.

72. C. Papadimitriou and K. Steiglitz, *Combinatorial Optimization: Algorithms and Complexity*. Englewood Cliffs, NJ: Prentice Hall, 1982.

73. R. Jain, *The Art of Computer Systems Performance Analysis: Techniques for Experimental Design, Measurement, Simulation and Modeling*. New York: Wiley, 1991.

74. J. Jia, J. Zhang, and Q. Zhang, "Cooperative relay for cognitive radio networks," *Proc. IEEE INFOCOM*, pp. 2304–2312, April 2009.

75. Y-C. Liang, K. C. Chen, G. Y. Li, and P. Mahonen, "Cognitive radio networking and communications: An overview," *IEEE Transactions on Vehicular Technology*, vol. 60, no. 7, pp. 3386–3407, September 2011.

76. D. Willkomm, S. Machiraju, J. Bolot, and A. Wolisz, "Primary users in cellular networks: A large-scale measurement study," *Proc. IEEE International Symposium on New Frontiers in Dynamic Spectrum Access Networks*, pp. 1–11, October 2008.

77. M. Fitch, M. Nekovee, S. Kawade, K. Briggs, and R. Mackenzie, "Wireless service provision in TV white space with cognitive radio technology: A telecom operator's perspective and experience," *IEEE Communication Magazine*, vol. 49, no. 3, pp. 64–73, May 2011.

78. J. Yao, S. Kanhere, and M. Hassan, "Improving QoS in high-speed mobility using bandwidth maps," *IEEE Transactions on Mobile Computing*, vol. 11, no. 4, pp. 603–617, April 2012.

79. R. Murty, R. Chandra, T. Moscibroda, and P. Bahl, "SenseLess: A database-driven white spaces network," *IEEE Transactions on Mobile Computing*, vol. 11, no. 2, pp. 189–203, February 2012.

80. W. Wang and X. Liu, "List-coloring based channel allocation for open-spectrum wireless networks," *Proc. IEEE Vehicular Technology Conference*, pp. 690–694, September 2005.

81. H. Zheng and C. Peng, "Collaboration and fairness in opportunistic spectrum access," *Proc. IEEE International Conference on Communications*, pp. 3132–3136, May 2005.

82. K. Yang and X. Wang, "Cross-layer network planning for multi-radio multi-channel cognitive wireless networks," *IEEE Transactions on Communications*, vol. 56, no. 10, pp. 1705–1714, October 2008.

83. C. Perez-Vega and J. M. Zamanillo, "Path-loss model for broadcasting applications and outdoor communication systems in the VHF and UHF bands," *IEEE Transactions on Broadcasting*, vol. 48, no. 2, pp. 91–96, June 2002.

84. C. Perez-Vega and J. L. Garcia, "Frequency behavior of a power-law path loss model," *Proc. Microcoll*, Budapest, March 1999.

85. Federal Communications Commission, "Code of Federal Regulations," Title 47. Ch. 1, Part 73, Radio Broadcast Services. Secs. 73.683, 73.684, and 73.699.

86. Theodore S. Rappaport, *Wireless Communications: Principles & Practice*. Englewood Cliffs, NJ: Prentice Hall, 1996.

87. M. Hata, "Empirical formula for propagation loss in land mobile radio services," *IEEE Transactions on Vehicular Technology*, vol. 29, no. 3, pp. 317–325, 1980.

88. K. Yang, N. Prasad, and X. Wang, "An auction approach to resource allocation in uplink OFDMA systems," *IEEE Transactions on Signal Processing*, vol. 57, no. 11, pp. 4482–4496, November 2009.

89. R. E. Burkard and E. Cela, "Linear assignment problems and extensions," *COMPUTG: Computing (Archive for Inf. Numer. Computat.)*, January 1998.

90. R. Jain, D. Chiu, and W. Hawe, "A quantitative measure of fairness and discrimination for resource allocation in shared computer systems," *DEC Research Report TR-301*, September 1984.

91. C. Gellings, "The concept of demand-side management for electric utilities," *Proceedings of the IEEE*, vol. 73, no. 10, pp. 1468–1470, October 1985.

92. H. Farhangi, "The path of the smart grid," *IEEE Power and Energy Magazine*, vol. 8, no. 1, pp. 18–28, January 2010.

93. U.S. Department of Energy, "The smart grid: An introduction," Technical Report, 2008.

94. W. Wang, Y. Xu, and M. Khanna, "A survey on the communication architectures in smart grid," *Computer Networks*, vol. 55, no. 15, pp. 3604–3629, October 2011.

95. X. Fang, S. Misra, G. Xue, and D. Yang, "Smart grid—The new and improved power grid: A survey," *IEEE Communications Surveys & Tutorials*, vol. 14, no. 4, pp. 944–980, December 2012.

96. O. Fatemieh, R. Chandra, and C. A. Gunter, "Low cost and secure smart meter communications using the TV white spaces," *Proc. International Symposium on Resilient Control Systems*, pp. 37–42, August 2010.

97. A. Ghassemi, S. Bavarian, and L. Lampe, "Cognitive radio for smart grid communications," *Proc. IEEE International Conference on Smart Grid Communications*, pp. 297–302, October 2010.

98. M. Brew, F. Darbari, L. H. Crockett, M. B. Waddell, M. Fitch, S. Weiss, and R. W. Stewart, "UHF white space network for rural smart grid communications," *Proc. IEEE International Conference on Smart Grid Communications*, pp. 138–142, October 2011.

99. K. Nagothu, B. Kelley, M. Jamshidi, and A. Rajaee, "Persistent net—AMI for microgrid infrastructure using cognitive radio on cloud data centers," *IEEE Systems Journal,* vol. 6, no. 1, pp. 4–15, March 2012.

100. R. Yu, Y. Zhang, S. Gjessing, C. Yuen, S. Xie, and M. Guizani, "Cognitive-radio-based hierarchical communications infrastructure for smart grid," *IEEE Network,* vol. 25, no. 5, pp. 6–14, September 2011.

101. Q. D. Vo, J.-P. Choi, H. M. Chang, and W. C. Lee, "Green perspective cognitive radio-based M2M communications for smart meters," *Proc. International Conference on Information and Communication Technology Convergence*, pp. 382–383, November 2010.

102. S. Gong and H. Li, "Dynamic spectrum allocation for power load prediction via wireless metering in smart grid," *Proc. Annual Conference on Information Sciences and Systems*, pp. 1–6, November 2011.

103. M. Fahrioglu and F. Alvarado, "Using utility information to calibrate customer demand management behavior models," *IEEE Transactions on Power Systems*, vol. 16, no. 2, pp. 317–322, May 2001.

104. S. H. A. Ahmad, M. Liu, T. Javidi, Q. Zhao, and B. Krishnamachari, "Optimality of myopic sensing in multichannel opportunistic access," *IEEE Transactions on Information Theory*, vol. 55, no. 9, pp. 4040–4050, September 2009.

105. H. Kim and K. G. Shin, "Efficient discovery of spectrum opportunities with MAC-layer sensing in cognitive radio networks," *IEEE Transactions on Mobile Computing*, vol. 7, no. 5, pp. 533–545, May 2008.

106. T. Jiang, Z. Wang, L. Zhang, D. Qu, and Y-C. Liang, "Efficient spectrum utilization on TV band for cognitive radio-based high-speed vehicle network," *IEEE Transactions on Wireless Communications*, vol. 13, no. 10, pp. 5319–5329, October 2014.

107. T. Jiang, Y. Cao, L. Yu, and Z. Wang, "Load shaping strategy based on energy storage and dynamic pricing in smart grid," *IEEE Transactions on Smart Grid*, vol. 5, no. 6, pp. 2868–2876, November 2014.

Index

A

Access point, 105
ACL; *See* Available channel list
Adaptive collaboration sensing scheme
　　(ACSS), 41–46
　　assignment matrix, 43
　　basic idea, 41–42
　　brute-force search, 44
　　channel states, 42
　　decision rule for SPRT, 43
　　false alarm, probability of, 43
　　finite state variables optimization
　　　problem, 44
　　log-likelihood ratio, 42
　　optimal sensing node allocation,
　　　43–46
　　performance, 51
　　principle of optimality, 44
　　pseudo-code (dynamic
　　　programming), 46
　　sequential probability ratio test,
　　　42–43
　　Viterbi algorithm, 44
Adaptive opportunistic spectrum access
　　(AOSA) scheme, 31
Additive white Gaussian noise
　　(AWGN), 38
Advanced meter infrastructure (AMI)
　　communications, 12

Auction algorithm, 91
Available channel list (ACL), 82–83

B

Base load generator (BLG), 101
Base stations (BSs)
　　bottleneck, 8
　　cellular, 10
　　cognitive, 38, 79
　　deployment, 57
　　half-duplex cooperation, 9
　　primary, 76
Baum-Welch algorithm (BWA), 26
Bayes' theorem, 106
Broadband wireless communications
　　(BWC), 10, 75

C

CBS; *See* Cognitive base station
CDF; *See* Cumulative distribution
　　function
Cellular communication, 8–10,
　　57–73
　　base stations, 8
　　bottleneck link between the base
　　　station and cellular user, 8
　　BS-to-RS/CU and RS-to-CU
　　　transmissions, 9
　　cellular users, 8

cognitive radio-assisted
cooperation framework, 58–64
BS operations, 60
cellular network, 59
channel side information, 59
CU operations, 61
DBS allocation constraints, 63
decode-and-forward protocol, 59
dedicated band sub-channels, 58
direct transmission, 61, 62
downlink data streams, 60
full-duplex cooperation, 61, 62
half-duplex cooperation, 61, 62
power constraints, 64
PU network, 59
RS operations, 60
white space sub-channels, 58
WSS allocation constraints, 64
cooperative relaying, 8
downlink transmissions, 57
half-duplex cooperation, 9
joint resource allocation, 10
multiple-in multiple-out-based
cellular networks, 8
network coding assisted
cooperation scheme, 8
network sum utility, 8
optimal transmission power
allocation, 64–68
CRAC problem, 66
cross-layer optimization, 64–65
Hungarian algorithm, 68
Lagrange multiplier vector, 64
power constraint elimination,
65–66
sub-gradient method, 65
throughput maximization, 66–68
orthogonal frequency division
multiple access cellular
networks, 8
performance analysis and
evaluation, 68–73
COST-231 model, 68
fairness index, 69
impact of cell population, 70

impact of primary user traffic
load, 70–73
line-of-sight component, 68
Markov ON-OFF process, 68
network utility, 69
performance comparisons, 69
simulation scenario, 68–69
throughput gain, 69
preassigned sub-channels, 8
relay stations, 8
target system, 9
Cellular users (CUs)
direct communication to, 58
quality of service for, 8, 57
Central limit theorem, 39
Cognitive base station (CBS), 38, 76, 78
Cognitive radio networks, introduction
to, 1–14
broadband wireless
communications, 13
cellular communications, 8–10
base stations, 8
bottleneck link between the base
station and cellular user, 8
BS-to-RS/CU and RS-to-CU
transmissions, 9
cellular users, 8
cooperative relaying, 8
half-duplex cooperation, 9
joint resource allocation, 10
multiple-in multiple-out-based
cellular networks, 8
network coding assisted
cooperation scheme, 8
network sum utility, 8
orthogonal frequency division
multiple access cellular
networks, 8
preassigned sub-channels, 8
relay stations, 8
target system, 9
cognitive radio-based networks,
1–4
coining of term, 2
definition, 3

key words, 3
legacy command-control
    regulation, 1
licensed spectrum, utilization of,
    2
model-based reasoning, 3
radio knowledge representation
    language, 3
resource allocation schemes, 4
content and organization, 13–14
cooperative sensing, 6–7
    idle channel, 7
    machine-learning-based
        multi-band spectrum sensing
        policy, 7
    MAC-layer sensing, 6
    Markov decision process, 6
    sensing sequence, 7
    sequential probability ratio test, 6
    sequential sensing algorithms, 7
high-speed vehicles, 10–11
    broadband wireless
        communications, 10
    distributed greedy algorithm, 11
    graph-theoretical model, 11
    large-scale combinatorial
        optimization problem, 11
    non-deterministic polynomial
        complexity, 11
    primary channel occupancy
        modeling, 10
    primary users, 10
    wireless local area networks, 10
opportunistic spectrum access
    networks, 4–6
    constrained Markov decision
        process, 5
    energy constraint, 5
    interference, 5, 6
    multi-channel systems, channel
        selection in, 5
    secondary user transmission
        scheme, 5
smart grid, 11–12
    advanced meter infrastructure
        communications, 12

advantages, 11
dynamic pricing, 11
fundamental issue, 12
Industrial Scientific Medical
    unlicensed spectrum bands,
    11–12
load shaping methods, 11
multi-objective genetic
    algorithms, 12
spectrum scarcity, 12
Cooperative sensing, 6–7, 37–55
    adaptive collaboration sensing
        scheme, 41–46
    assignment matrix, 43
    basic idea, 41–42
    brute-force search, 44
    channel states, 42
    decision rule for SPRT, 43
    false alarm, probability of, 43
    finite state variables optimization
        problem, 44
    log-likelihood ratio, 42
    optimal sensing node allocation,
        43–46
    principle of optimality, 44
    pseudo-code (dynamic
        programming), 46
    sequential probability ratio test,
        42–43
    Viterbi algorithm, 44
cumulative distribution function, 55
idle channel, 7
machine-learning-based multi-band
    spectrum sensing policy, 7
MAC-layer sensing, 6
Markov decision process, 6
multi-user multi-channel system
    model, 38–41
    additive white Gaussian noise,
        38
    central limit theorem, 39
    cognitive base station, 38
    energy detector, 39
    fusion center, 40
    goal of spectrum sensing, 38
    idle channels, 40

licensed spectrum band, 38
MAC-layer sensing framework,
    38
null hypothesis, 39
Nyquist theorem, 39
Rayleigh fading, 38
received signal-noise-ratio, 39
sequential probability ratio test
    theory, 38
performance evaluation and
    analysis, 47–52
conventional schemes, 50
fast spectrum acquisition, 48, 51
Monte Carlo searching, 50
Okumura-Hata propagation
    model, 47
Rayleigh fading, 47
Rayleigh fading channel,
    probability density function,
    54
sensing delay, 37
sensing sequence, 7
sequential probability ratio test, 6
sequential sensing algorithms, 7
COST-231 model, 68
Cumulative distribution function (CDF),
    20, 55
CUs; *See* Cellular users

D
Decode-and-forward (DF) protocol, 59
Dedicated band sub-channels (DBSs),
    58, 63
Downlink
BS operations, 60
OFDMA scheme, 8
resource allocation, 10
transmissions, 57

E
Electricity load shaping framework
    (smart grid), 100–106
Element number of the minimum
    independent vehicle set
    (ENMIVS), 93

Energy
constraint, 5
cost, 99, 107
detection, 6, 39
storage, 101, 113, 115

F
Fairness index, 69, 96
Fast spectrum acquisition, 48, 51
FCC; *See* U.S. Federal Communications
    Commission

G
Geolocation Database, 78

H
Hidden Markov model (HMM), 26, 30
High-speed vehicle network, 10–11,
    75–98
broadband wireless
    communications, 10, 75
cognitive radio-based high-speed
    vehicle network, 76–83
available channel list, 82–83
cognitive base station, 76
coverage radius, 76
Geolocation Database, 78
low-speed vehicles, 77
Okumura-Hata propagation
    prediction model, 82
path loss, 81
poss loss model, 80–82
primary base station, 76
relative primary channel
    occupancy model, 78
slot-based spectrum handoff, 78
spectrum sharing list, 83
spectrum utilization, 79
system model, 76–79
TV broadcasting in VHF and
    UHF bands, 81
TV protection range, 80
TV signal activity model, 78
TV signal transmission tower, 76
vehicle communication range, 80

distributed greedy algorithm, 11
graph-theoretical model, 11
large-scale combinatorial
    optimization problem, 11
maximization of utilized white
    space, 83–92
    Auction algorithm, 91
    branch and bound search
        method, 87–89
    Hungarian algorithm, 91
    integer programming problem
        with a linear constraint, 85
    Karush-Kuhn-Tucker conditions,
        90
    linear programming method with
        low complexity, 90–92
    relationship matrix, 87
    separation computing, 85–87
    single-channel method with low
        complexity, 89
    truncate algorithm, 91
non-deterministic polynomial
    complexity, 11
performance analysis and
    evaluation, 92–98
    element number of minimum
        independent vehicle set, 93
    Jain's fairness index, 96
    Monte Carlo simulations, 93
    randomly deployed PBSs, 92
    TV signal transmission tower,
        93
primary channel occupancy
    modeling, 10
primary users, 10
wireless local area networks, 10, 75
HMM; See Hidden Markov model
Hungarian algorithm, 68, 91
Hyper-Erlang distribution, 22–23

I
Industrial Scientific Medical (ISM)
    unlicensed spectrum bands,
    11–12, 100

Integer programming problem with a
    linear constraint (IPLC), 85

J
Jain's fairness index, 69, 96

K
Karush-Kuhn-Tucker (KKT) conditions,
    90
Keep-on allocation strategy, 106

L
Lagrange multiplier vector, 64
Line-of-sight (LOS) component, 68
Load differentiation dynamic pricing
    (LDDP), 102
Log-likelihood ratio (LLR), 42

M
Markov channel model, 104
Markov ON-OFF process, 68
Multiple-in multiple-out (MIMO)-based
    cellular networks, 8
Multi-user multi-channel system model
    (cooperative sensing), 38–41
    additive white Gaussian noise, 38
    central limit theorem, 39
    cognitive base station, 38
    energy detector, 39
    fusion center, 40
    goal of spectrum sensing, 38
    idle channels, 40
    licensed spectrum band, 38
    MAC-layer sensing framework, 38
    null hypothesis, 39
    Nyquist theorem, 39
    Rayleigh fading, 38
    received signal-noise-ratio, 39
    sequential probability ratio test
        theory, 38

N
Non-deterministic polynomial (NP)
    complexity, 11
Null hypothesis, 39
Nyquist theorem, 39

O

OFDMA; *See* Orthogonal frequency
    division multiple access
Okumura-Hata propagation model, 47,
    82
ON/OFF model, 16
Opportunistic spectrum access network,
    4–6, 15–36
    constrained Markov decision
      process, 5
    energy constraint, 5
    hidden Markov model, 16
    interference, 5, 6
    multi-channel systems, channel
      selection in, 5
    optimal probabilistic slot
      allocation, 21–23
      baseline performance, 21
      exponential distribution, 21–22
      hyper-Erlang distribution, 22–23
      Poisson arrival traffic model, 21
    performance analysis and
      evaluation, 23–34
      adaptive opportunistic spectrum
        access scheme, 31
      baseline scheme, 27
      Baum-Welch algorithm, 26
      collision probability, 24
      false alarm events, 24
      false alarm probability, 25
      hidden Markov model, 26
      idle channel, 23
      impact of sensing errors, 23–25
      impact of unknown primary user
        idle period distribution, 26–27
      missed detection events, 23
      performance comparisons, 27–34
      sensing errors, 25, 30
    probabilistic slot allocation
      scheme, 18–20
      probability density function, 20
      sensing sub-plot, 19
      spectrum hole, 18, 20
      transmission probabilities, 19

secondary user transmission
    scheme, 5
single-user single-channel system
    model, 16–18
    half-duplex transceiver, 16
    interference constraint, 16
    ON/OFF model, 16
    spectrum holes, 16
    transmission structure, 17
Optimality, principle of, 44
Orthogonal frequency division multiple
    access (OFDMA), 8
    cognitive radio-assisted
      cooperation framework, 59
    direct transmission, 61
    downlink transmissions, 57
    frame structure, 60
    spectrum balancing problem, 64
    sub-channels, 68

P

PDF; *See* Probability density function
Peak clipping, 11, 99
Peak load generator (PLG), 101
Poisson arrival traffic model, 21
Primary base station (PBS), 76
Primary user (PU), 10, 16
Probabilistic slot allocation (PSA)
    scheme, 18–20
    probability density function, 20
    sensing sub-plot, 19
    spectrum hole, 18, 20
    transmission probabilities, 19
Probability density function (PDF)
    hyper-Erlang distribution, 22
    probabilistic slot allocation
      scheme, 20
    Rayleigh fading channel, 54
    sequential probability ratio test, 42

R

Radio knowledge representation
    language (RKRL), 3
Random allocation strategy, 106

Rayleigh fading, 38, 47, 54
Relationship matrix, 87
Relative primary channel occupancy
    model, 78
Relay stations (RSs), 8
    allocation, 10, 13
    bottleneck, 8
    half-duplex cooperation, 9
    sub-channel, 58

S
Secondary user (SU), 16
Sensing channel allocation; *See* Smart
    grid
Sensing node allocation; *See*
    Cooperative sensing
Sequential probability ratio test (SPRT),
    38
    cooperative sensing, 6
    decision rule, 42, 43
    multi-user multi-channel system
        model, 38
    properties, 43, 44
    thresholds for, 44
Signal-noise-ratio (SNR), 39, 50, 106
    increasing, 50
    Rayleigh fading channel, 54
    received, 39, 106
Slot-based communication, 105
Smart grid, 11–12, 99–116
    advanced meter infrastructure
        communications, 12
    advantages, 11
    characteristic, 99
    dynamic pricing, 11
    electricity load shaping framework,
        100–106
        access point, 105
        base load generator, 101
        Bayes' theorem, 106
        clustered communication
            architecture, 105
        cognitive radio network model,
            104–106
        communication errors, 106

electricity load shaping
    framework, 101
    energy detection, 106
    idle channels, discovery of, 105
    inelastic electricity, 100
    load differentiation dynamic
        pricing, 102
    Markov channel model, 104
    peak load generator, 101
    pricing strategy, 101
    received signal-to-noise ratio,
        106
    slot-based communication, 105
    smart grid model, 100–103
fundamental issue, 12
Industrial Scientific Medical
    unlicensed spectrum bands,
        11–12, 100
load shaping methods, 11, 99
multi-objective genetic algorithms,
    12
peak clipping, 11, 99
performance analysis and
    evaluation, 109–115
    communication reliability, 109,
        110
    electricity load shaping, 111–115
    idle channels, 114
    Monte Carlo searching, 111
    pseudo-code of obtaining the
        approximate optimal, 110
    sensing channel allocation,
        109–111
    total electricity load over time
        slots, 112
sensing channel allocation and load
    shaping strategies, 106–109
    action policy, 108
    experience function of consumer,
        107
    grid electricity price, 107
    inelastic electricity demand, 108
    Keep-on allocation strategy, 106
    load shaping strategy, 107–109
    random allocation strategy, 106

sensing channel allocation
  strategies, 106–107
spectrum scarcity, 12, 100
spectrum sensing, accuracy of, 116
SNR; *See* Signal-noise-ratio
Spectrum sharing list (SSL), 83
SPRT; *See* Sequential probability ratio
  test
Stanford Graphics software, 81
SU; *See* Secondary user

T
Transmission power allocation; *See*
  Cellular communication
Transmission slot allocation; *See*
  Opportunistic spectrum access
  network
TV signal(s)
  activity model, 78

path loss, 81
transmission tower, 76, 93

U
UHF bands, broadcasting in, 81
Uplink transmission, 58
U.S. Federal Communications
  Commission (FCC), 1

V
VHF bands, broadcasting in, 81
Viterbi algorithm, 44

W
White space
  allocation; *See* High-speed vehicle
    network
  sub-channels (WSSs), 58, 64
Wireless local area network (WLAN),
  10, 75